Impact assessment and sustainable resource management

THEMES IN RESOURCE MANAGEMENT
Edited by Professor Bruce Mitchell, University of Waterloo

Already published:
John Chapman: Geography and Energy
R. L. Heathcote: Arid Lands: their use and abuse
Adrian McDonald and David Kay: Water Resources: issues and strategies
Peter H. Omara-Ojungu: Resource Management in Developing Countries
John T. Pierce: The Food Resource

Forthcoming:
D. S. Slocombe: Managing Resources for Sustainability

L Graham Smith

Impact Assessment and Sustainable Resource Management

Longman Scientific & Technical

Copublished in the United States with
John Wiley & Sons, Inc., New York

Longman Scientific & Technical,
Longman Group UK Ltd,
Longman House, Burnt Mill, Harlow,
Essex CM20 2JE, England
and Associated Companies throughout the world.

Copublished in the United States with
John Wiley & Sons, Inc., 605 Third Avenue, New York, NY 10158

ISBN 0 582 046 53X

First published 1993

British Library Cataloguing in Publication Data
A catalogue record for this title is available from the British Library.

Library of Congress Cataloging-in-Publication Data
A catalogue record for this title is available from the Library of Congress.

Set in Linotron Times 10/12pt by 8 II

Printed in Malaysia by VVP

To Christine

Contents

Contents

List of figures

Acknowledgements

A number of people have made significant contributions to the completion of this book. All of the figures were designed by Trish Chalk. They were produced by Trish and Gord Shields at the Department of Geography's Cartographic Centre, the University of Western Ontario. Their assistance and ability to translate ideas into graphic images are much appreciated. Much of the work of the book was completed while on sabbatical leave from the Department of Geography at the University of Western Ontario. The support of the department, the university and a Social Sciences and Humanities Research Council travel grant is appreciated.

As Series Editor, Bruce Mitchell provided constructive criticism, advice and guidance in his usual calm and professional manner. Throughout the writing process Bruce was supportive and he was a steady source of inspiration. His friendship and counsel are truly valued.

At Longman, Vanessa Lawrence took a personal interest in the book and the well-being of its author, despite an increasingly busy workload and hectic travel schedule. Her support and guidance were invaluable. The staff at Longman endeavoured to keep the production of the book on schedule. Their patience and perseverance are much appreciated.

Intellectually, a debt is owed to those who have helped shape many of the ideas explored in the text. In particular, the influence of Ralph Krueger, Gordon Nelson, Mike Goodchild and the late Derrick Sewell, is much appreciated. In addition, Dan Shrubsole, Tom Carlisle and Lars Hurlen all read various parts of the text and gave valuable comments on how it could be improved. On differing occasions, discussions with Keith Storey, Bob Gibson, Chris Wood, John Horberry and Tony Dorcey all helped provide insight and direction for the text. Lastly, the book benefits greatly from the reactions, comments and experiences of students at the University of Western Ontario since 1985. Their efforts and willingness to experiment with differing teaching formats were a direct influence on the book and its structure.

Acknowledgements

I should like to thank Larson for creating *The Far Side*. It and the companionship of our cats (Tyson, Pudkins, Simon, Jesse and Poco) helped me to keep the activities of humans in perspective while writing the book.

I owe most to my family. Both the Smiths in Britain and the Knights in Canada are a constant source of love and support. This book is dedicated to my wife, Christine. She patiently keeps the true meaning of life in focus. Both Courtney and Joanna arrived while this book was being written. Together, they keep the smile on my face and the warmth of love in my daily life. Thank you.

Graham Smith
Shedden, Ontario
March, 1992

Foreword

The Themes in Resource Management Series has several objectives. One is to identify and to examine substantive and enduring resource management and development problems. Attention will range from local to international scales, from developed to developing nations, from the public to the private sector and from biophysical to political considerations.

A second objective is to assess response to these management and development problems in a variety of world regions. Several responses are of particular interest but especially *research* and *action programmes*. The former involves the different types of analysis which have been generated by natural resource problems. The series will assess the kinds of problems being defined by investigators, the nature and adequacy of evidence being assembled, the kinds of interpretations and arguments being presented, the contributions to improving theoretical understanding as well as resolving pressing problems, and the areas in which progress and frustration are being experienced. The latter response involves the policies, programmes and projects being conceived and implemented to tackle complex and difficult problems. The series is concerned with reviewing their adequacy and effectiveness.

A third objective is to explore the way in which resource analysis, management and development might be made more complementary to one another. Too often analysts and managers go their separate ways. A good part of the blame for this situation must lie with the analysts who too frequently ignore or neglect the concerns of managers, unduly emphasize method and technique, and exclude explicit consideration of the management implications of their research. It is hoped that this series will demonstrate that research analysis can contribute both to the development of theory and to the resolution of important societal problems.

Graham Smith's book is the tenth in the Themes in Resource Management Series. *Impact Assessment and Sustainable Resource Management* explores some of the fundamental issues associated with impact assessment, identifies current strengths and weaknesses, and suggests what changes are needed if impact assessment is to contribute to the achievement of sustainable resource management. Smith suggests that impact assessment represents as much a philosophy and a process as it does a technique, and that to date too much attention has been focused on technique questions. He argues that if significant improvement is to occur the time has come to consider basic issues related to the philosophy and the process of impact assessment.

As with other books in the series, Smith establishes systematic framework for problem solving, and then illustrates concepts and methods with examples from various world regions. The reader will find case studies from countries such as the United States, Canada, Britain, Germany, Switzerland, Portugal, Brazil, Senegal and Bangladesh. In addition, the activities associated with organizations such as the World Bank and initiatives such as Superfund in the United States are examined.

Graham Smith received his undergraduate degree at the University of Leeds, his master's degree at the University of Victoria in British Columbia and his doctorate at the University of Waterloo. In his work, he has focused upon policy analysis, impact assessment and public participation, with particular attention to energy planning and management.

As a result of his work over the past decade, he has had many opportunities to deal directly with the types of problems and issues which he addresses in his book. Thus, his ideas and suggestions have emerged both from extensive reading in the international literature and from having to deal with conceptual and methodological problems in research and consulting situations. The outcome is a book which balances both academic and practical considerations. It therefore should be of value to readers with a variety of interests in impact assessment.

Bruce Mitchell
University of Waterloo
Waterloo, Ontario

March 1992

CHAPTER 1

The need for redefinition

Introduction

Impact assessment needs to be redefined. Despite two decades of evolution and a myriad of techniques, present practice appears unable to help in preventing environmental disasters such as the *Exxon Valdez* oil-tanker spill off Alaska, or poor resource management such as the wholesale removal of tropical rain forest in the Amazon or the demise of species such as the giant panda in China. These problems do not arise out of ignorance. They have not occurred because developments were unplanned, nor their impacts unforeseen. Rather, they are the results of a flawed conceptualization of impact assessment and its role in environmental planning and resource management.

Too often, impact assessment has been viewed as a hurdle in the path of resource development, or as the means by which development proposals can be justified and environmental objections appeased. Environmental protection has been viewed as a desirable but distinctly secondary objective within resource management. Moreover, environmental planning has often been viewed as an entity independent of human activity. The development, extraction and exploitation of the resource base have taken precedence over the assurance of long-term environmental sustainability. The goal of sustainable resource management is to redress this imbalance.

This book develops the concept of impact assessment as *a process for resource management and environmental planning that provides for the achievement of the goal of sustainability*. In this first chapter, the basic tenets of sustainability are examined. Sustainable resource management requires an approach to decision making that is integrative, adaptive and interactive. In view of these requirements, the chapter outlines traditional approaches to decision making in resource management and the factors that led to the emergence of impact assessment. It is argued that impact assessment in its present form has not realized its full potential and that the need exists for impact assessment to be redefined.

1

Sustainability

The concept of sustainability originated with the 1980 *World Conservation Strategy* of the International Union for the Conservation of Nature and Natural Resources (IUCN). The IUCN advanced sustainability as a strategic approach to the integration of conservation and development consistent with the objectives of:

- ecosystem maintenance
- the preservation of genetic diversity,
- sustainable utilization of resources.

This served as the antecedent to further promotion of the concept of *sustainable development* by the World Commission on Environment and Development, established in 1983 by the United Nations to formulate 'a global agenda for change'. The Commission was headed by Gro Harlem Brundtland, Prime Minister of Norway, and published its final report, *Our Common Future*, in 1987.

Sustainable development was defined in a general manner by the World Commission on Environment and Development (1987: 43) as 'development that meets the needs of the present without compromising the ability of future generations to meet their own needs'. The key aspects of sustainable development relate to an understanding of:

- *environment*: as not just the biophysical, natural domain but also the socio-political, human components that constitute a global environment for which there is an interdependent, world ecology

- *development*: as not just an economic activity but as a process of qualitative and equitable growth

- *society*: as an interdependent, world community reliant upon a single biosphere wherein global economic growth cannot succeed with an uneven distribution of wealth

- *linkages*: among poverty, inequality and environmental degradation.

The central idea is that development can occur only if and when there is recognition of the need to sustain and expand the environmental resource base. The associated corollary is that 'economic growth, in and of itself, is insufficient for the purposes of development' (Shearman, 1990: 6).

There is a substantial and growing literature on the topic of sustainability (e.g. Clark and Munn, 1986; Brown et al, 1987; Jacobs and Munro, 1987; Redclift, 1987; Rees, 1988; Turner, 1988a; Archibugi and Nijkamp, 1989; Daly, 1990; Dovers, 1990; Pearce and Turner, 1990; Shearman, 1990; Rees, 1990a). Much of this literature has been concerned with the meaning and application of sustainability within the field of environmental economics and the distinctions between 'sustainability', 'sustainable development', 'sustainable utilization' and 'sustainable growth'. In general, there seems to be little disagreement that sustainability arguments stress the need for environmental protection and continuing economic growth to be viewed as mutually compatible rather than conflicting objectives (Turner, 1988b: 5). The area of confusion appears to be in providing a precise definition of what sustainability will involve in practice, particularly from an economic standpoint. O'Riordan (1988: 30) has suggested that sustainability in the purest sense involves 'embracing ethical norms pertaining to the survival of living matter, to the rights of future generations and to institutions responsible for ensuring that such rights are fully taken into account in policies and actions'. In contrast, sustainable utilization and sustainable growth are more 'manageable and politically acceptable' manifestations of the sustainability concept.

Citing a need for a more explicit definition of sustainable development, Brown et al (1987) expressed a concern that sustainability was evolving into a 'transcendent term', subject to frequent but imprecise usage. In contrast, Shearman (1990: 1) presented a compelling argument based on the premise that 'it is not sustainability that requires definition or clarification, but rather its implications for any given context to which it is applied'. Shearman maintained that sustainability is used as a modifier as in sustainable development, sustainable growth, sustainable ecosystems, etc. and that it is more important to understand the implicative meaning of sustainability within the context that it is used. Thus, rather than attempting to develop precise definitions of sustainability, Shearman (1990: 3) advocated using sustainability as a concept and focusing debate on the issues implied by sustainability rather than the issue of sustainability.

As a concept, sustainability implies that there is an inherent contradiction in the pursuit of the basic goal of development through a reliance upon approaches to economic growth that may instead actually result in human suffering (Shearman, 1990; Redclift, 1987; Sen, 1984). For many, growth is synonymous with increasing wealth. However, as Daly (1990: 5) has surmised: 'What is in dispute is whether growth, at the current margin, is really making us wealthier. As growth in the physical dimensions of the human economy pushes beyond the optimal scale relative to the biosphere it in fact makes us poorer. Growth, like anything else, can cost more than it is worth at the margin.'

Sustainability is a response to the growing recognition that several, seemingly endemic, global problems can no longer be divorced from a consideration of a threatened future. Present patterns of resource distribution, gross national product (GNP), industrial power, trade and transnational corporations all act to reinforce an uneven distribution of development (Pirages, 1989). This imbalance is mirrored by the prevailing pattern of world inequality indicated by measures of poverty, debt, starvation and out-migration (Kidron and Segal, 1984). These realities all reinforce a dominant

North–South pattern that characterizes what Frank (1966) termed the 'development of under-development' (Jones, 1983; Independent Commission on International Development Issues, 1980; Pirages, 1978; Wallerstein, 1979).

The concern is not simply one of short-term development and prosperity versus poverty, but involves deeper issues of distribution and allocation relative to the future (Pirages, 1989). Moreover, because all communities and societies must share the same earth with each other whatever their differences and inequalities, the threats posed by environmental stress, uncontrolled growth and environmental impacts in different parts of the globe are threats that affect the future of the whole globe.

Poverty can generate a misuse of resources. Immediate survival often promotes an emphasis upon exploitation and excessive consumption of the local resource base that, in the longer term, only serves to build external dependencies. Environmental stress leaves areas more prone to natural disasters, exacerbates heavy reliance upon resource exports, and intensifies dependency on world markets (World Commission on Environment and Development, 1987: 67–89; Jacobs and Munro, 1987).

Modern technology has permitted an era of unprecedented growth. Indeed, since 1950 there has been an explosion in world population, urbanization, industrialization and economies. Unfortunately, there has been a parallel increase in famine and pollution. For many, there is now a sense of technology beyond human control:

> Humanity is . . . entering an era of chronic, large-scale, and extremely complex *syndromes* of interdependence between the global economy and the world environment. Relative to earlier generations of problems, these emerging syndromes are characterized by profound scientific ignorance, enormous decision costs, and time and space scales that transcend those of most institutions. (Clark, 1986: 5)

The resulting environmental impacts potentially are irreparable. Examples include the greenhouse effect, ozone depletion, acid rain, global climate change, marine pollution, toxic wastes, desertification and the loss of forests. Not only do these activities result in environmental impacts in the form of species eradication, habitat loss and the reduction of ecological diversity and resiliency, but they also contribute to world economic crises, inflation, debt and starvation (Clark and Munn, 1986; Rees, 1990b).

Whereas these various concerns are significant individually, the tendency is for them to occur in combination. It is clear that:

- Environmental stresses are linked, making individual, *ad hoc* problem solving insufficient.
- Stress and the pattern of development are linked, which means that global gaps and inequalities are continuing to grow.
- Environmental and economic problems are a function of social and

political factors, rendering purely technical solutions to environmental problems inadequate.
- These systemic factors operate both within and between nations, such that unilateral solutions are not likely to be effective.

In a lucid and succinct manner, the Brundtland Report summarized this situation and called for a reconsideration of future decision making based on a balanced attention to environment, development and society: 'In essence, sustainable development is a process of change in which the exploitation of resources, the direction of investments, the orientation of technological development, and institutional change are all in harmony and enhance both current and future potential to meet human needs and aspirations' (World Commission on Environment and Development, 1987: 46).

Sustainability is best viewed as a concept. It is a value-based concern that requires 'the moral choice of accepting intergenerational equity as an overriding ethic' (Dovers, 1990: 299). To achieve that ethic, a sustainable future would display several key attributes (Fig. 1.1). Sustainability is a social goal, which by its very nature, takes the form of 'an over-riding, often ill-defined, but all-pervasive directive' (Dovers, 1990: 299). The challenge, therefore, is not to develop a precise definition for sustainability but to develop 'a conceptual framework for addressing issues in sustainability in order to understand and appreciate what would be involved in cultivating and initiating appropriate environmental planning and policy' (Shearman, 1990: 7)

Sustainability rests on the tenet that 'technology and social organization can be both managed and improved to make way for a new era of economic growth' (World Commission on Environment and Development, 1987: 8). The key is in *how to manage* technology and social organizations in resource development to provide for decision making that will foster sustainability.

Traditional decision making

Until the 1950s, resource management decision making mostly addressed the following questions:

- Is it technically feasible?
- Is it financially viable?
- Is it legally permissible?

Not surprisingly, this rudimentary form of evaluation often resulted in engineering-based answers to resource management problems, the 'technical fix' usually involving some form of structural solution, chosen from a narrow range of predetermined options. Provided the scale of development was limited, the problem well defined and adequate information available, this system of resource decision making functioned quite well. However, the increasing complexity, scale and implications of resource development

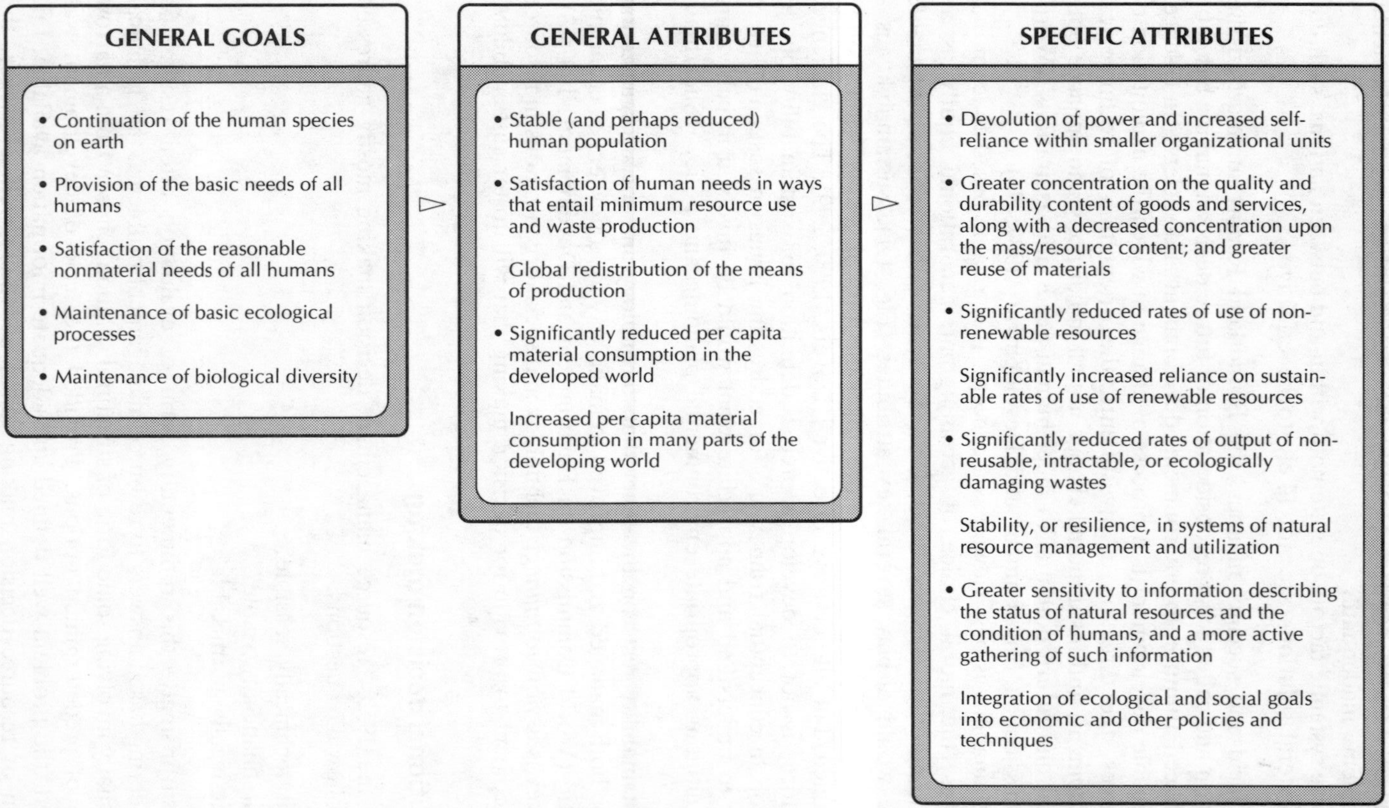

GENERAL GOALS	GENERAL ATTRIBUTES	SPECIFIC ATTRIBUTES
• Continuation of the human species on earth	• Stable (and perhaps reduced) human population	• Devolution of power and increased self-reliance within smaller organizational units
• Provision of the basic needs of all humans	• Satisfaction of human needs in ways that entail minimum resource use and waste production	• Greater concentration on the quality and durability content of goods and services, along with a decreased concentration upon the mass/resource content; and greater reuse of materials
• Satisfaction of the reasonable nonmaterial needs of all humans	• Global redistribution of the means of production	• Significantly reduced rates of use of non-renewable resources
• Maintenance of basic ecological processes	• Significantly reduced per capita material consumption in the developed world	• Significantly increased reliance on sustainable rates of use of renewable resources
• Maintenance of biological diversity	• Increased per capita material consumption in many parts of the developing world	• Significantly reduced rates of output of non-reusable, intractable, or ecologically damaging wastes
		Stability, or resilience, in systems of natural resource management and utilization
		• Greater sensitivity to information describing the status of natural resources and the condition of humans, and a more active gathering of such information
		Integration of ecological and social goals into economic and other policies and techniques

Fig. 1.1 Attributes of sustainability
Source after Dovers (1990)

projects through the 1950s revealed the weaknesses of this approach. Often there was more reliance upon (hidden) political agendas and ulterior motives than any sound technical scrutiny of projects. Many projects resulted in major environmental degradation. Moreover, the wider social costs of such developments as dam construction and flood control schemes were shown to be largely unaccounted despite their significance (White, 1945, 1957; Maass, 1951). The desire for a broader form of social accounting within resource decisions led to the advent of benefit–cost analysis.

Benefit–cost analysis emerged initially in the evaluation of water resource developments in the United States as a means of broadening the approach to evaluation in resource management decision making (Sewell, 1973). Through the 1960s, benefit–cost analysis became the predominant technique for evaluation in resource decision making, its popularity a function of the simplicity with which its objectives could be understood and the apparent ease with which it could be applied to a wide range of situations.

The merits of benefit–cost analysis and the principles of its application have been well outlined (e.g. Sewell et al, 1965; Mishan, 1976). Briefly, benefit–cost analysis may be used to:

- assess the economic characteristics of a particular project
- determine which of several projects designed to serve a given purpose results in the largest ratio of benefits to costs
- determine which of several projects designed to serve different purposes confers the largest net benefit on the economy as a whole.

Benefit–cost analysis was attractive to engineers and decision makers because it could produce a tangible measure of 'social utility', generating quantitative indices relating fiscal benefits and costs, with an emphasis on the criterion of economic efficiency.

However, the technique also suffered from a surfeit of misapplication in practice, including: a failure to consider alternatives; a focus on easily measured, quantifiable benefits and costs; a failure to adhere to key premises, leading to inflated benefit measures and manipulated accounting; and an inability to account accurately for, and incorporate, such temporal changes as variations in interest rates, adequate discount rates and price levels. In addition, benefit–cost analysis was criticized for its conceptual inabilities to account for the distributional aspects of costs and benefits and the problems of aggregation (Carley and Bustelo, 1984: 139–49).

These deficiencies prompted the development of alternative means for the economic appraisal of projects, such as the goals achievement matrix of Hill or the planning balance sheet developed by Lichfield (McAllister, 1980). Simple benefit–cost analysis was replaced by more sophisticated variants, using multiple objectives and discount rates, proxy-pricing mechanisms and various means of planning, programme budgeting and cost-effectiveness analysis. However, this increasing sophistication could not avoid criticisms that centred on the inappropriateness of such accounting procedures for the evaluation of complex environmental interrelationships and the broader social issues of resource allocation (O'Riordan and Sewell, 1981).

In summary, benefit–cost analysis was (and is) a good technique when applied properly and when its limitations are recognized. Too often this has not been the case and a vociferous, adversarial debate developed concerning not only the methodological aspects of economic evaluations but also the political and ethical dimensions of their use (Swartzman et al, 1982).

Thus, while benefit–cost analysis remained the pre-eminent technique for resource decision making, the need for alternatives became apparent. The desire for an alternative form of social accounting was further reinforced by two interrelated factors: the increasing scale, complexity and uncertainty associated with resource development proposals, and the growth of public opposition to the approval of those projects. The era of the megaproject had arrived, but it was accompanied by the rebirth of public activism, particularly in the case of environmental quality and the desire for equity in the processes of governance. The outcome of these assorted pressures was the appearance of environmental impact assessment.

The evolution of impact assessment

Environmental impact assessment (EIA) originated with the passage of the National Environmental Policy Act (NEPA) in the United States in 1969: NEPA established the requirement for an environmental impact statement (EIS) as the principal means of implementing impact assessments. This 'enduring legacy' not only symbolized a new commitment to environmental protection but was an 'affirmation of faith' in the use of science for planning and decision making (Sadler, 1986: 102). In this respect, NEPA reflected a society 'mesmerized by the benefits of economic growth and dominated by a technocratic perspective on problem solving' (Boothroyd and Rees, 1984: 2). The emergence of EIA was as much bounded by, as it was a reaction to, this reality.

There were three components to NEPA (Ortolano, 1984: 139–51):

- *A policy statement*: sections 101A and 101B contain a declaration of national policy on the environment and the 'continuing responsibility' of the federal government for environmental matters.
- *A set of 'action forcing' provisions*: requiring that federal agencies must
  · utilize an interdisciplinary approach to planning (section 102:2:A)
  · develop procedures to give environmental factors 'appropriate consideration' in decision making (section 102:2:B)
  · prepare EISs (section 102:2:C)
- *The creation of the Council on Environmental Quality (CEQ)*: to assist the President in preparing an annual report on environmental quality, appraise federal agency performance in implementing NEPA, conduct research and advise.

These requirements were a significant departure from the traditional approach to decision making in resource management and were a response to the demand for proper environmental design of large resource developments (O'Riordan and Sewell, 1981).

Section 102:2:C was the most crucial within NEPA. It required that all agencies of the federal government:

> include in every recommendation or report on proposals for legislation and other major Federal actions significantly affecting the quality of the human environment, a detailed statement by the responsible official on
> (i) the environmental impact of the proposed action,
> (ii) any adverse environmental effects which cannot be avoided should the proposal be implemented,
> (iii) alternatives to the proposed action,
> (iv) the relationship between local short-term uses of man's environment and the maintenance and enhancement of long-term productivity, and
> (v) any irreversible and irretrievable commitments of resources which would be involved in the proposed action should it be implemented.

This clause demarcated the EIS as *the* focus for impact assessment and, under NEPA, a model for EIA emerged that was product driven, with scientific data collection preceding positivistic analysis and the production of technical reports. Moreover, it was this NEPA-based model that set the standard for the establishment of impact assessment processes elsewhere.

In practice, this model of impact assessment was often ineffective (O'Riordan, 1981). Difficulties included:

- A general lack of adequate data bases.
- The time horizon allotted for studies was too short for thorough investigation.
- There was a marked absence of socially related data.
- The weighting of findings remained problematical.

Further criticisms of NEPA arose through the flood of subsequent litigation that served both to interpret the practical meaning of the statute and to underscore the fact that most agencies were merely tacking on EIA to predetermined decisions; only reviewing 'proximate alternatives' rather than fundamental choices (Fairfax and Ingram, 1981; Friesma and Culhane, 1976; Fairfax, 1978).

Planning reactions to the perceived failings of NEPA involved attempts to improve the science of impact analysis by making impact statements more analytical, readable and informative, but not reconceived in approach. Consequently, EIA was still shunted aside on important issues in favour of 'hard' data from the engineering design and economic feasibility components of proposals. Moreover, impact assessment continued to exhibit a distinct biophysical bias. Social systems were often not understood and there was minimal to no consideration of social impacts in most impact statements.

By the late 1970s it was apparent that EIA had fallen short of its promise and a transition stage ensued with the efficacy of impact assessment the focus of attention (Boothroyd and Rees, 1984). It was clear that the science of EIA had not been good, but it was equally evident that the problem was not merely one of scientific training. For example, as Boothroyd and Rees (1984, 4) stated: 'The ecologists who had become involved may have had a better sense of the complex dynamics of natural systems, but could offer no coherent theory of ecosystem behaviour under stress.' Moreover, the ability of EIA to protect the environment by influencing the approvals process was brought into question by a paralysing quantity of NEPA related litigation in the United States. This propensity for litigation has characterized the perceived failure of NEPA. It can also be seen as a portent of the manner with which impact assessment initiatives can be effectively 'frozen' by a failure to consider adequately the implications they pose to the existing institutional arrangements for management (Boothroyd and Rees, 1984; Fenge and Smith, 1986; Smith, 1988a; Wandesforde-Smith and Kerbavaz, 1988).

The result of this period of reflection and review of the initial experience was an expansion of the concept of impact assessment. In the 1980s impact assessment ceased to be merely 'environmental' by title with the advent of social impact assessment, community impact assessment, technology assessment, risk analysis and adaptive environmental assessment and management. This titular proliferation reflected a practical realization and frustration that impact assessment had not reached its potential. However, attempts to resolve these concerns through refinements to the technical design elements of impact assessment are misguided: as misguided as the earlier efforts to refine benefit–cost analysis for environmental analysis.

The predominant rationale in developing various forms of impact assessment has been a concern for the poor level and quality of science within existing impact assessments. The paucity of good science (pure, applied and/or social) is perceived to be operating within well-defined administrative procedures and, as one major review noted, the 'result often has been a somewhat confused and frustrating technical review process' (Beanlands and Duinker, 1983: 1). This situation is seen as the principal cause of frustration in denigrating the ability of science to achieve a mandate for environmental protection within impact assessment. The solution usually proposed is the refinement of scientific concepts to define, specify and quantify better the design and conduct of assessment studies.

For example, in their attempt to determine the extent to which the science of ecology could be used to refine the design and conduct of EIA, Beanlands and Duinker (1983: 2) proposed that 'significant improvements' in the scientific quality of assessments could be realized through the reduction of constraints involving the need for:

- a common standard for an acceptable scientific basis to impact assessment
- early agreement on the approach to be adopted
- continuity of study such that impact assessment does not cease with the production of an EIS
- information transfer,
- better communications.

However, rather than confronting these limitations of impact assessment directly, Beanlands and Duinker adopted the premise that 'the institutional framework for environmental impact assessment is in place before the scientific basis has been established' (Munn, as cited in Beanlands and Duinker, 1983: 13) and confined themselves to a consideration of the ways by which the adoption of ecological principles, and ecology as a science, would improve the practice of impact assessment *within those constraints*.

These constraints do not reflect a weakness of science. Rather, they reflect the reality that impact assessment has evolved as an ongoing *political* process within development planning. In accepting the institutional arrangements premise as fixed, proponents of impact assessment derivatives can only refine the technical aspects of the impact assessment process within the very constraints they themselves acknowledge as delineating the ability of impact assessment to achieve full fruition. As much as this line of research might advance the veracity of impact assessment studies, it is inherently constrained from improving in any fundamental fashion the conceptual basis of impact assessment itself. Improving the science of environmental analysis *per se* does nothing to reform the political processes of resource management that govern how that information is utilised. An alternative response is warranted: one that necessitates redefining the role of impact assessment.

A new role for impact assessment

Mitchell (1990b: 15) suggested that after an initial period of adoption, two options exist regarding the implementation of innovations and new concepts in resource management: 'we need to consider whether the task is to improve our capability to implement the concept, or whether we should re-examine the foundations of the concept itself'. As the preceding discussion has shown, the emphasis in impact assessment, thus far, has centred on attempts to improve the implementation of the concept. In contrast, the central tenet of this book is the need to re-examine the foundations of impact assessment itself.

This book seeks to redefine the role of impact assessment within resource management. The basis for that redefinition is developed in Chapters 2–5. Chapter 2 reinforces the basic theme of the book and reviews the present status of impact assessment methods and methodology. In Chapter 3, the institutional arrangements for impact assessment are discussed and a framework for their analysis detailed. Chapter 4 examines public policy and interest representation, emphasizing the nature of public policy as a process involving interest representation, decision making and administration. Planning and the role of impact assessment provide the focus for Chapter 5. This chapter examines the meaning of planning and details a normative model for the integration of impact assessment into environmental planning.

A new role for impact assessment is then defined within Chapter 6. The new definition rests on the postulates that:

- Impact assessment should be designed as a bridge that integrates the science of environmental analysis with the politics of resource management.
- The institutional arrangements for decision making are central to the definition of the role played by impact assessment in resource management.
- A model for sustainable resource management can provide the means for integration.

Chapter 6 presents a renewed definition of impact assessment and details its role within *an integrative framework for sustainable resource management*.

Finally, the utility of this new conceptualization is illustrated through the examination of various case studies drawn from the international experience with impact assessment for frontier developments (Ch. 7), linear facilities (Ch. 8), and waste management (Ch. 9). A final and tenth chapter presents some concluding observations and suggested directions for future research.

CHAPTER 2

Impact assessment methods and methodology

Introduction

Impact assessment originated with a desire for profound change in both the philosophy and the methodology of resource management. Common to all forms of impact assessment is 'an assumption that a systematic, focused, interdisciplinary use of science may improve the quality of planning and decision-making' (Caldwell, 1988a: 75). This tenet first found expression with the passage in the United States of the National Environmental Policy Act (NEPA) in 1969, reviewed in Chapter 1.

Writing in 1988, one of the architects of NEPA, Lynton Caldwell, reflected that the intent of impact assessment was a reform in the 'priorities affecting the environment and in procedures for administrative decision-making' (Caldwell, 1988a, 75). Not only did NEPA provide the enabling legislation for that reform, but it also required a significant change in traditional agency practice because it dictated the 'interdisciplinary use of the natural and social sciences and the environmental design arts' to address non-quantified values in the employment of ecological information and concepts (Caldwell, 1988a: 76–7).

Impact assessment represented a considerable challenge to traditional decision making and was met with scepticism. The implementation of impact assessment entailed changes neither incremental in design nor superficial in formulation. How would this change be manifest? How would agencies adjust to this new requirement? What form would impact assessment take *in practice*?

The answers to these questions reveal much about the realization of impact assessment and the extent to which the ideals of early proponents have survived the bureaucracy and politics of policy implementation. This chapter outlines the nature of impact assessment as it has evolved since the early 1970s. It suggests that the development of impact assessment has focused upon methodological exigency at the expense of philosophical principle.

Types of impact assessment

Early versions of impact assessment were always prefaced by the adjective 'environmental'; as in environmental impact analysis, environmental assessment or environmental impact assessment. As the field developed, additional prefixes surfaced such as technology assessment, social impact assessment and community assessment. These titles reflect differences in subject emphasis (the biophysical environment, technology, human communities); geographic focus (local, regional, national); time horizon (immediate, the next generation, the distant future); format (project statements, planning scenarios, investment plans); and uses of the output (project approval, policy appraisal, plan determination), of the various types of impact assessment (Rossini and Porter, 1983: 11).

However, all forms of impact assessment ideally have several key features in common (Rossini and Porter, 1983: 4–5). All impact assessments should be:

- effects-focused
- future oriented
- centred around technological developments
- systematic, comprehensive and interdisciplinary in approach,
- comparative and policy-oriented.

Based on these commonalities, three major types of impact assessment can be recognized:

- *technology assessment* (TA)
- *environmental impact assessment* (EIA)
- *social impact assessment* (SIA).

Technology assessment is the most general form of impact assessment. It is best understood as a category of public policy analysis that focuses upon the prospective societal consequences of changes in technology (Coates and Coates, 1989). Technology assessments are open-ended searches, usually in the form of 'a policy study that assesses all significant impacts of the use of a new technology' (Finsterbusch, 1980: 21). In the United States, this function has been institutionalized by the creation of the Office of Technology Assessment of Congress (see Coates and Coates, 1989). Elsewhere, the attention given to technology assessment is less formally structured and it is not as clearly defined as a significant sub-field of impact assessment.

Environmental impact assessment remains strongly associated in the United States with the requirements of NEPA. Traditionally, EIA has emphasized the biophysical sciences, providing a process whereby the principles and paradigms of ecological science can be applied to the consideration of the ecological effects of various forms of development on natural ecosystems (Hirst, 1984). Elsewhere, this biophysical emphasis may be more implicit than explicit and EIA is often perceived as being synonymous with the only form of impact assessment.

Social impact assessment stresses the impacts of technological development, environmental modification and/or planned interventions on human communities. In the United States SIA is closely associated with the work of the Army Corps of Engineers and, in both the United States and elsewhere, it has evolved as the principal offshoot from EIA, with its own extensive literature (Finsterbusch and Wolf, 1977; Finsterbusch, 1980; Finsterbusch et al, 1983; Leistritz and Murdock, 1981; Tester and Mykes, 1981).

Carley and Bustelo (1984) noted that most SIAs have been directed at resource development and/or large scale projects. They also stressed the importance of the political and institutional context within which SIA operates as a determinant of its likely effectiveness. In this respect, the basic principles for SIA include the recognition of pluralism in political decision making and the need for public participation in impact assessment (Carley and Bustelo, 1984). Clearly, these dimensions of the decision process apply to all forms of impact assessment and, in general, a prominent feature of the different forms of impact assessment 'is the presence of many issues of common concern in their theoretical underpinnings, methodologies and substantive attention' (Cope and Hills, 1988: 175).

However, in exploring the need for *total assessment*, Cope and Hills (1988: 184) cautioned that: 'Whatever the origins of the ever-increasing demands for assessment and the ever-widening notions of which topics are relevant to incorporate in assessment procedures, it remains true that the assessment process as such is founded on conceptual principles which do not embrace all the components of human action and motivation.' Impact assessment originated with a desire to broaden the disciplinary base of environmental decision making. It should not be overly compartmentalized nor divided into separate sub-fields. Interdisciplinary analysis is an implicit component of impact assessment and requires that *environment* be understood in all its dimensions: as much social, technological and cultural as biophysical. Similarly, effective resource management necessitates that a broad approach be applied to environmental evaluation and that an *integrative approach* be adopted: one that uses a multiplicity of means and strategies, blends various resource sectors, provides a mechanism for social and economic change, and seeks to achieve accommodation and compromise between the various interests (Mitchell,1986; Lang, 1986a). Thus, while acknowledging that several distinct variations in emphasis can exist, this book adopts the convention of *impact assessment* and advocates the view that an integrative perspective is crucial to the full realization of the aims and purposes of impact assessment.

Definitions of impact assessment

Impact assessment began as an alternative means to benefit–cost analysis for social accounting. Consistent with these origins, a dominant paradigm for impact assessment has emerged which consigns impact assessment to the limited role of a technique to improve the decision making data base for project implementation. For example, impact assessment has been defined as referring to:

a process or set of activities designed to contribute pertinent environmental information to project or programme decision making.
(Beanlands and Duinker, 1983: 18)

a process which attempts to identify, predict and assess the likely consequences of proposed development activities. (Munro et al, 1986: 2)

a planning aid . . . concerned with identifying, predicting and assessing impacts arising from proposed activities such as policies, programmes, plans and development projects which may affect the environment. (Bisset, 1983a: 131)

a basic tool for the sound assessment of development proposals . . . to determine the potential environmental, social and health effects of a proposed development. (Clark, 1983a: 4)

These definitions present a consistent view of impact assessment as *a technique* to improve the data base for decision making through a process of *information generation* related to the identification, prediction and assessment of the effects of project implementation. This view limits impact assessment in two key respects: first, it confines impact assessment to a relatively narrow focus, its role being that of information generation; and second, it places the emphasis upon methodology, with a priority on the need to develop and identify the ways and means by which:

- potential impacts of projects can be identified
- their likely effects predicted,
- their consequences determined.

Methodology

Various methods for impact assessment were developed because administrative provisions, by themselves, were insufficient to ensure that an impact statement would contain structured information, produced in a 'scientific' manner, useful for decision making (Bisset, 1983a). Impact assessment must deal with four key problem areas (Jain et al, 1977; Bisset, 1987):

- *Identification*: identifies likely effects and determines the degree of comprehensiveness and specificity to be used in isolating project impacts, their timing and duration. What data sources are to be employed?
- *Prediction and measurement*: explicit indicators of impact must be determined and the magnitude of those impacts estimated. Concerns regarding uncertainty and the relative objectivity and subjectivity of the data must be addressed.

- *Interpretation*: explicit criteria must be used to determine the significance of impacts, to indicate the level of uncertainty and risk involved, and to compare alternatives. Should the impacts be aggregated or disaggregated? Is there to be any public involvement in the interpretation of impacts?
- *Communication*: results must be communicated effectively to affected parties and should contain a setting description and a summary format. Are the key issues clear? Does the communication satisfy regulatory compliance?

In addressing these considerations, the process for impact assessment involves three steps (Carley and Bustelo, 1984):

- the establishment of a data base to describe the existing situation
- the development of the means to describe changes related to the project
- the forecasting of changes in the base situation with and without the given project, including both qualitative and quantitative aspects.

The first of these steps involves techniques for data collection in both the biophysical and the social sciences. A wide range of these techniques exist. For example, both Ortolano (1984) and Westman (1985) have reviewed physical scientific techniques in environmental studies and impact assessment, various environmental engineering approaches have been summarized by Rau and Wooten (1980), and Finsterbusch et al (1983) have outlined social scientific means for data acquisition in impact assessment. Many of these techniques have standardized procedures and protocols to guide their application. But few, if any, of these procedures are unique to impact assessment; nor should they be confused with methods for impact assessment. Rather, they simply represent the application of existing environmental science within the realm of impact assessment.

This distinction has not always been clear and there has been a surfeit of attention within impact assessment to the collection and analysis of *baseline data* – information that describes the environmental conditions prior to the advent of the activity necessitating impact assessment. One estimate indicated that 'baseline-type surveys and inventories accounted for 50–90% of total time, effort and budget' expended on impact assessments (Hirst, 1984: 205).

Normally, baseline studies emphasize the collection of quantitative data. If they are properly designed, baseline studies can function as the foundation for long-term monitoring of carefully selected ecosystem parameters. Proper design entails data collection in a statistically valid manner, with specific research hypotheses in mind. Unfortunately, the practice in many impact assessments has been in stark contrast to this ideal: 'The so-called "shotgun" approach has prevailed, with comprehensive but superficial coverage of all elements of the environment, regardless of their relevance to project decisions' (Beanlands and Duinker, 1983: 2). Moreover, the term 'baseline studies' has been employed as a catchall phrase to include the entire range of pre-project studies. Unfortunately, the studies are normally limited to descriptive, one-time surveys of all the various components of the environment. Seldom is it clear what the objectives are, what limitations there are on data interpretation or what use is made of the results. (Beanlands and Duinker, 1983: 29). Volumes of baseline data, regardless of their accuracy, precision or means of acquisition, do not, in and of themselves, constitute an impact assessment. The key aspects, therefore, are the latter two steps of change description and forecasting: the elements of impact identification, prediction and evaluation.

Five principal types of method have been used to perform the tasks of impact identification and summarization (see Bisset, 1983a, b, 1987, 1988; Canter, 1983; Hyman and Stiftel, 1988; Jain et al, 1977; Shopley and Fuggle, 1984; Thompson, 1990; Wathern, 1988a; Westman, 1985; Whitney and Maclaren, 1985). In order of increasing sophistication, they are:

- *checklists*
- *interaction matrices*
- *overlay mapping*
- *networks*
- *simulation modelling*.

Checklists

Checklists are standard lists of features which may be affected by a project. They represent the simplest form of approach to assessing project impacts and were one of the earliest methods for impact assessment to be developed. Checklists aim to promote thinking about impacts, providing a concise summary of the effects of proposals, identifying factors and the trade-offs between alternatives. There are several basic formats for checklists (Bisset, 1987), including:

- *Simple lists*: a listing of potentially affected factors. The list acts to focus attention on these attributes and the checklist acts as a guide to impact identification. No guidelines or information are included on how the various factors are to be measured.
- *Descriptive checklists*: measurements and predictive techniques for each factor are included in the checklist to provide guidance on the assessment of the impacts identified by the listing. This results in a more adequate method of data collection, with both the potential impact and its constituent elements being considered.
- *Scaling checklists*: criteria for evaluation are incorporated into the listing, usually in the form of a subjective rating or scaling. The checklist is set up as a worksheet with space to indicate the relative significance of each impact, along with critical values which represent a 'threshold of concern' for each factor.
- *Multi-attribute utility theory*: in an extension of scaling checklists, a

subjective measure of relative merit (a utility function) is derived for each of the environmental parameters listed in the checklist. The impacts of various project alternatives are then compared on the basis of these derived values.

Some of the more well-known and widely used checklist variants are well described by Hyman and Stiftel (1988: 155–223) and by Canter (1983). These include the wetland evaluation technique (WET), a descriptive checklist developed for the US Army Corps of Engineers; the water resources assessment methodology (WRAM), a scaled checklist also developed by the US Army Corps of Engineers; and, the environmental evaluation system (EES) developed by Dee and others at the Batelle Columbus Laboratories (Dee et al, 1973).

The EES is a scaled checklist which assigns scores (value functions) relating to the impact on each of 78 parameters relating to ecology, environmental pollution, aesthetics and human interests. These scores are then transformed into a single, overall value representing the predicted impact for each project alternative. Standardized graphs are used to perform these data transfers and the key idea behind the EES approach is to identify the parameters most sensitive to impact as a result of the proposed project.

Similar to the EES, the Sondheim method is also a scaled checklist producing a single measure for the projected impact of each alternative (Sondheim, 1978). The Sondheim method differs from the EES, however, in two key aspects. First, the significance attached to each individual parameter in the 'weighting' of the overall impact of an alternative is project-specific rather than a predetermined standard. Second, the process of aggregation is not solely reliant upon scientific expertise, as is the case with the EES, but utilizes public involvement.

Lastly, decision analysis (Keeney, 1980; Keeney and Raiffa, 1976) represents an example of the final variant of checklists, multi-attribute utility theory. This approach is complex, but it does incorporate an explicit consideration of probability and sensitivity analysis (Bisset, 1988). Decision analysis proceeds through a defined series of steps (Hyman and Stiftel, 1988):

- The problem is structured through the identification of objectives and the definition of measurement or 'attributes' for each of the objectives.
- Future values for each attribute are predicted by a team of experts.
- The experts indicate a preference ranking for each attribute and derive an estimate of relative merit (the utility function) on the basis of this preference and the expected future values.
- Mathematical optimization procedures are applied by the analyst to the derived utility functions to determine which of the various alternatives for a project would maximize expected utility.

Despite its potential for application within impact assessment, because of its emphasis upon the role of multiple objectives, decision analysis has not been used to any great extent. This might be a function of its mathematical complexity and its explicit reliance upon value judgements. However, decision analysis also uses utility as a common measurement unit. A problem arises from the fact that while utility may be a widely accepted concept in economic theory, it 'is much more widely accepted as a theoretical rather than as a practical tool for analysis' (Whitney and Maclaren, 1985: 29). Thus, while it may have theoretical clarity, decision analysis lacks the ease of understanding that is a pragmatic necessity for any approach to impact assessment. It is a method 'designed primarily for applications with a single identifiable decision maker' (Hyman and Stiftel, 1988: 212) and it is ill-suited to the public decision making characteristic of most impact assessment situations.

Whichever particular format is employed, checklists are a good technique for structuring the initial stages of an assessment and for guidance on which alternatives should be considered. However, as the sole method for impact assessment, checklists have a number of limitations. They have a rigid format and tend to foster tunnel vision since they are based upon a conception of impact that is limited to the changes in a single parameter. In addition, checklists can be generalized, do not show interactions, may double count and may end up as immense lists.

Interaction matrices

Interaction matrices were developed from a desire to link environmental factors with project activities. Matrices are grid diagrams with one set of factors on the horizontal axis and another on the vertical. The interaction between components on the opposing axes is recorded in the cell common to both in either a presentational manner, using symbols or numerical scores, or in a mathematical manner, using algebraic functions (Shopley and Fuggle, 1984).

The most well-known example of an interaction matrix is that developed by Leopold et al (1971). The Leopold matrix is intended to reveal two-dimensional relationships between impacts and actions. It provides a systematic checking of each development activity against a listing of environmental factors. The matrix lists 100 possible project actions (such as modification of habitat, urbanization, surface excavation, etc.) within 10 general categories on the horizontal axis and lists 88 environmental factors (such as soils, flora, land-use, etc.) within 4 general categories on the vertical axis.

Whereas the full matrix has a total of 8800 cells, not all activities nor the full range of environmental factors will apply in each instance. To use the matrix, the analyst first checks all of the activities that may be associated with a project. Next, all of the environmental factors are examined and a slash placed in those cells where an impact is possible for each activity that is marked. A 10-point scale is then utilized to score levels of impact relative to both their magnitude and their importance. In this manner, the matrix serves both as a checklist and as a summary of the impact assessment (Hyman and Stiftel, 1988: 167–72).

The Leopold matrix is strictly a means for organizing an impact analysis. No measurement strategies for the projected impacts are prescribed (the matrix relies upon the evaluations of experts), nor is there any indication of how (or if) the various magnitude and impact scores should be weighted (Hyman and Stiftel, 1988: 170). Indeed, the Leopold matrix retains many of the weaknesses identified for checklists and can be viewed as being 'only a series of checklists for different actions' (Westman, 1985: 138). In addition, the size of the full matrix (8800 cells) can make its use time-consuming and its results difficult to comprehend. There are numerous opportunities for double counting within the matrix, it has a heavy biophysical bias and there are limited opportunities to reveal secondary effects (Westman, 1985: 138).

Not all matrices have these same weaknesses and interaction matrices can serve a valuable role in impact identification and in the presentation of comparative information for different project alternatives (Canter, 1983). The matrix approach is also valuable in situations where an absence of baseline studies and other inhibiting conditions may preclude other approaches to impact assessment. Thus, they have been shown to have great utility within developing countries as a means to provide initial environmental examinations and as a tool for 'rapid impact assessments' (Lohani and Halim, 1987).

Overlay mapping

Recognizing the spatial nature of many environmental impacts, overlay mapping was pioneered by McHarg (1969). It is an approach based on the principles of land capability mapping. The method was first used manually. A base map of the area has superimposed upon it transparent sheets summarizing each set of features of the existing environment according to value classes connoting high (dark shading), medium (light shading) or low (no shading) quality for that particular element of the environment. Impacts are shown through the density of shading which builds up as each attribute is overlaid.

Although simple in its conception and application, the McHarg approach to overlay mapping provides a powerful visual representation of potential impacts wherein the spatial distribution of project effects can be seen and the results easily comprehended by affected publics. It is a method particularly well suited to the assessment of impacts associated with alternative routes for linear facilities such as highways, transmission lines and pipelines.

The disadvantages to the McHarg approach to overlay mapping are part technical and part conceptual. From a technical standpoint, the manual superimposition of more than 12 overlays is problematic (Bisset, 1987). From a conceptual standpoint, the McHarg approach can be accused of environmental determinism since it is primarily concerned with classifying land and infers impacts based upon an assumption that all environmental factors are equally weighted (Hyman and Stiftel, 1988: 162: Westman, 1985: 227–52). Moreover, while overlays are excellent for portraying the spatial character of impacts, they indicate little about impact probability or duration. Overlays cannot reflect synergistic effects arising from the interaction of two or more factors. Causal relationships are not addressed, nor do they distinguish between direct and indirect effects.

The technical restraints on overlay mapping were largely overcome through the use of computers to assist both with data storage and in the analysis of the resulting land use patterns. Early examples of computer-interactive approaches to overlay mapping include the metropolitan landscape planning model (METLAND), centred on the Boston region, and the SIRO-PLAN programme, developed in Australia by the Land Use Research Division of CSIRO, the Commonwealth Scientific and Industrial Research Organization (Westman, 1985: 242–50). Another major area of development in computer-assisted mapping has been in the field of corridor siting for linear facilities. A good illustration is the application of 'constraint mapping' in the identification of preferred routes for high-voltage transmission lines in Ontario (Smith and Cattrysse, 1987).

Further development of the overlay approach is likely as impact assessment makes better use of geographical information systems (GIS). Because both impact duration and causal relationships can be addressed through the application of GIS technology, the advent of GIS has already gone a long way to resolving the conceptual problems inherent in earlier applications of overlay mapping. Moreover, because a GIS encodes spatial information in digital format, it makes those data available to mathematical analysis. This capability enables a GIS to deal with two other major problems in overlay mapping:

- *tract heterogeneity*: ensuring that the characterization of any mapped area is based upon attributes that are uniformly distributed and represented within that mapped area
- *boundary definition*: delineating the correct boundaries between areas of differing characterization.

A GIS provides not only a means for computer-assisted cartography but also a 'powerful set of tools for collecting, storing, retrieving at will, transforming, and displaying spatial data from the real world' in a manner that allows that data to be modelled and analysed (Burrough, 1986). In its simplest form, a GIS may be defined as 'an information system which has as its primary source of input, a base composed of data referenced by spatial or geographical coordinates' (Estes et al, 1987: 359). Because the data are stored in digital form, they may be retrieved, subjected to advanced mathematical transformation and analysis through computerized modelling and optimization techniques, and the results displayed visually through computer-assisted cartography. Moreover, through the use of remote sensing technology and satellite imagery (see Curran, 1987), the data within a GIS may be updated on a regular basis and impact predictions monitored against reality (Aronoff, 1989). Similarly, a GIS also provides a tool for the assessment of incremental changes in an environment and the determination of cumulative impacts (Johnston et al, 1988).

In summary, overlay mapping provides an excellent means to examine and visually display the spatial nature of impacts. It is also a useful method for planning linear developments, especially as a search mechanism for identifying and analysing alternative route preferences. Moreover, the continued advancement and development of GIS technology offer a means by which the main weaknesses of the method and its ability to deal with the issues of probability, time and causality, may be addressed. However, the widespread adoption of GIS is also problematical, as a GIS requires skilled personnel and access to appropriate hardware and software. Not all resource management agencies will have such capabilities.

Networks

Developed expressly to link secondary and tertiary impacts to primary impacts, networks are directional diagrams designed to trace in two dimensions the higher-order linkages between project actions and environmental factors. A network consists of a number of linked impacts known to have occurred in the past which are then used to trace 'the progression of causes and effects of various project actions' (Shopley and Fuggle, 1984: 39).

The strength of networks is their ability to identify and guide analysis to the indirect impacts which may arise from a project. They also have the capability to provide a visually understandable representation of those impacts. Networks do not contain criteria to determine impact significance and they are similar to other impact assessment methods in that they are still 'primarily a tool for identifying impacts, not evaluating them' (Hyman and Stiftel, 1988: 198). Networks also identify many more high order impacts than are likely to occur: differentiating those that will occur from those that will not usually requires more information than is available. Consequently, networks are rarely utilized, except in a highly abbreviated format, because of these informational constraints and the high costs implicit in their employment (Westman, 1985: 143–5; Hyman and Stiftel, 1988).

System diagrams are conceptually similar to networks in that they deal with linkages and depict 'environmental systems as complex arrangements of interrelated component parts' (Bisset, 1987: 47). Whereas networks attempt to show linkages between impacts, system diagrams link environmental components on the basis of energy flows between those components.

System diagrams, and their use in impact assessment, are based on the

23

work of Odum (1971) and the concept of energy as the fundamental unit of measurement within an ecosystem. Energy flows between environmental components thus represent a basic index of ecosystem functioning. These then can be depicted within a system diagram using symbols derived from electronic circuitry (Westman, 1985: 145–8; Bisset, 1987: 47–9).

Depicting ecological impacts in terms of energy flows, using a common unit of measurement (energy), is an attractive proposition. Unfortunately, this strength is offset by a number a major weaknesses that limit the usefulness of system diagrams for impact assessment. Within system diagrams, linked components are assumed to be more important than isolated elements, a potential bias exacerbated by the focus upon energy relations rather than ecological impacts (Westman, 1985). System diagrams, like networks, are also time-consuming and expensive to use. Lastly, system diagrams are exclusively biophysical in orientation and the inclusion of socio-economic aspects is 'fraught with conceptual and practical problems' (Bisset, 1988: 57).

Thus, both networks and system diagrams contain conceptual elements of value in the development of impact assessment methodologies. Impact assessment must seek to address higher-order impacts. The recognition of natural units of common measurement also merits further attention if assessment methods are to become more precise. However, in their existing format neither networks nor system diagrams lend themselves to widespread and frequent application.

Simulation modelling

The logical extension of networks is the application of mathematical and other sciences to the modelling of environmental systems. Certainly, an extensive array of mathematical, natural system and economic models is widely used within impact assessment studies (see for example the reviews of Basta and Bower, 1982; Biswas et al, 1990; de Broissia, 1986; Whitney, 1985). However, as a distinct methodology for impact assessment, simulation modelling involves the integration of standard modelling approaches in the natural and social sciences with the increased use and capabilities of computers to provide for differing impact scenarios to be developed, visualized and assessed.

Simulation models have three basic characteristics (Munn, 1983: 281). They are:

- simplified representations of the systems under investigation
- explicit assumptions regarding the behaviour of those systems,
- open to misinterpretation, especially if used out of context.

Models may be used to describe, explain and/or predict characteristics of environmental systems. Their greatest utility is in situations where there are few available data, considerable uncertainty as to the dynamic interrelationships between variables, and the simulation model is employed at an early investigative stage to aid in the conceptualization of the impact assessment study (Munn, 1983).

Examples of this kind of interactive simulation modelling include: the Kane simulation model (KISM), a computer algorithm for systems analysis applied to impact assessment; and, the habitat evaluation procedures (HEP), a widely used approach for the assessment of impacts upon fish and wildlife habitat developed by the US Fish and Wildlife Service (Hyman and Stiftel, 1988: 198–218). However, the most comprehensive example of simulation modelling for impact assessment is the adaptive environmental assessment and management (AEAM) approach based upon the work of Holling (1978; see also ESSA, 1982; Everitt, 1983; Jones and Greig, 1985; Walters, 1986).

Holling (1978) intended that adaptive environmental assessment and management would differ from conventional methods for impact assessment through the adoption of a broader focus and an emphasis upon coping with uncertainty rather than just improving impact predictability. His approach was developed to integrate environmental, social and economic understanding through the use of interactive simulation modelling. The AEAM approach attempts to deal with the data and prediction problems perceived in impact assessment practice. It uses small workshops to create a simulation model of the system(s) to be affected by development. Uncertainty is seen as a positive factor, rather than a negative one. Participants in the workshops must reach a consensus on the important features and relationships that characterize the system. Impacts are then modelled through an interactive computer simulation which allows variations to be tested and alternatives to be compared.

Discussing the application of AEAM in practice, Everitt (1983) stressed the value of simulation modelling in focusing or 'scoping' the impact assessment process. The AEAM approach is organized around the use of expert, interdisciplinary modelling workshops. This reliance upon workshops is considered by its proponents to be one of the principal strengths of AEAM as they address one of the main failings endemic to impact assessment techniques: poor communication (Jones and Greig, 1985).

The AEAM approach requires that the participants come together at the onset of a project to define collectively the terms of reference for the impact assessment. These participants are then involved through a continuing series of workshops in the computer simulation of the system they define as affected by, and as impacted upon by, development proposals (Jones and Greig, 1985).

The workshops occur early in the planning process. This not only allows impact predictions to be improved through the use of simulations but provides a forum for the expanded consideration of uncertainty (Everitt, 1983). So central is this explicit consideration of uncertainty that Jones and Greig (1985: 21) described AEAM as 'a collection of concepts and approaches whose common theme is the recognition that uncertainty is the dominant component of most environmental issues'. The AEAM approach is based on the motto of 'expect the unexpected' and proposes two ways to manage uncertainty: (1) reduce uncertainty through planning, and (2) develop projects capable of coping with uncertainty when it does occur (Jones and Greig, 1985).

As a method for impact assessment, AEAM has advantages in that it allows changes and effects to be seen quickly, allows changes in assumptions to be assessed and provides a useful management tool for operational projects. However, Bisset (1988: 58) has noted that there is a real need to broaden the range of experience with AEAM in practice before its full utility can be determined. Most of the literature on AEAM is by proponents of the method who have focused upon its procedural elements rather than the substantive contributions to improved impact assessment from the use of any simulations generated in practice. Moreover, it is not clear whether most agencies would be willing to invest in the time needed to run the AEAM procedure or if, indeed, they would have the necessary expertise to manage the complex logistics of the workshops upon which the method depends (Hyman and Stiftel, 1988: 221).

Despite these valid criticisms, it is clear that the AEAM approach to simulation modelling has had an influence on impact assessment thinking. As Bisset (1988) noted, AEAM has stimulated a re-evaluation of impact assessment methods and their implementation as regards their cost-effectiveness, predictive ability and accessibility to non-experts. The AEAM approach may not be perfect but it does underscore the need for impact assessment methods to be interactive, predictive and comprehensible.

Summary

No one methodology for impact assessment represents a panacea for all situations. All of the methods reviewed have in common a tendency to describe impacts rather than evaluate them. Some do not explicitly consider impact significance distinct from impact magnitude, few make provisions for public input to balance the judgements of 'experts' and all are project-specific. Thus, as with all research strategies, the principles of reliability and validity make it prudent for analysts to utilize more than one method in an investigation (Mitchell, 1989: 16–45). Moreover, in the case of impact assessment, there is a real paucity of empirical research that evaluates 'the actual operational performance of different types of method' for impact assessment (Bisset, 1988: 61).

Hence, any methodological recommendations are largely a function of supposition, personal preference and experiences that may, or may not, be directly applicable to a particular situation. In practice, the decision on which particular method should be used to conduct an impact assessment will vary with:

- the nature of the alternatives being assessed
- the role and degree of public participation desired
- resource availability
- familiarity
- issue significance,
- administrative constraints.

Complicating an analyst's choice of impact assessment methodology is the broadening range of situations and applications for which impact assessment approaches are now requested. Two examples that illustrate this point are cumulative effects assessment and risk assessment.

Cumulative effects assessment is concerned with impacts which (1) occur so frequently in time or so densely in space that they cannot be assimilated, or (2) combine with effects of other activities in a synergistic manner (Peterson et al, 1987; Sonntag et al, 1987). These impacts arise from a number of key activities within projects, notably:

- *Linear additive effects*: incremental additions to, or deletions from, a fixed storage where each increment has the same effect (e.g. toxic pollution of a lake).
- *Amplifying or exponential effects*: incremental additions to an apparently limitless storage where each increment has a larger effect than the one preceding (e.g. emissions of carbon dioxide into the global atmosphere).
- *Discontinuous effects*: incremental additions have no apparent consequence until a threshold is crossed and components change rapidly with distinctly different regimes of behaviour (e.g. lake eutrophication).
- *Structural surprises*: a function of multiple developments within a region, effects occur suddenly and locally. These effects slowly spread over a larger region and encompass not only ecological effects but an interdependence of economic, social and political mechanisms (e.g. the collapse of a fishery).

At issue is the point at which *impact saturation* may be said to have occurred: the threshold at which incremental activities, while in themselves seemingly benign, produce significant impacts.

Cumulative effects are not easily addressed within the normal, project-specific confines that characterize the dominant paradigm for impact assessment. Furthermore, it has been argued (Peterson et al, 1987; Sonntag et al, 1987) that the current conception of impact assessment:

- ignores the additive effects of repeated developments in the same ecosystem
- fails to address adequately precedent-setting developments that stimulate other activities
- ignores changes in the behaviour of ecological systems in response to increasing levels of perturbation,

27

- discourages the development of environmental objectives reflective of broader societal goals.

Cumulative effects pose a challenge to project-specific impact assessment. They are more easily understood within a wider land-use planning framework (Sebastiani et al, 1989) and are a further stimulus for a transition to a more comprehensive and holistic approach to impact assessment.

An environmental risk is a hazard or danger with adverse, probabilistic consequences. *Risk assessment* is concerned with how different societies, and elements within those societies, evaluate risk (Whyte and Burton, 1980; Kraft, 1988). Risk assessment involves three components:

- risk identification
- risk estimation,
- risk evaluation.

Impact assessment and risk assessment are mutually supportive concepts (Grima et al, 1989). More explicit attention to risk assessment will provide impact assessment with a means to address the associated issue of uncertainty and the role it plays within decisions (Grima et al, 1986; Fowle et al, 1988). Two questions are paramount: how do different segments of the public determine 'acceptable risk'?; and how should societal trade-offs be determined within public policy decisions?

These questions underscore the fact that while impact assessment 'continues to incorporate technical aspects of risk, there is a lot of scope for more emphasis on risk decisions as part of the social and political process' (Grima and Fowle, 1989: 155). Thus, risk assessment poses the challenge for impact assessment to become the means for broader societal planning that it was originally envisaged as being.

Both cumulative effects assessment and risk assessment reflect the failure of impact assessment to achieve its full potential within resource management. In most impact assessments, technical and economic factors are handled with clarity (usually numerically). Other factors, such as duration of impacts, uncertainty and public preferences, are usually less precise and controllable, remaining fuzzy and, ultimately, inconsequential to the final assessment of impact. The current status of impact assessment is one of good *technique*, but of poor *process*. In the words of Clark and Herington (1988: 12): 'an excessive interest in methodologies and techniques has . . . tended to direct attention away from viewing the experience of EIA within the broad process of environmental planning'.

Both in the United States and elsewhere, there has been undue emphasis upon the product of an impact assessment (the impact statement and its specific content) and, thus, by extension, upon methodology. There has been less attention to the need for good process in the design for, and management of, impact assessments:

We have now institutionalized a methodology for analyzing the

prospective impact of environment-altering propositions. Impact analysis has developed rapidly and widely as a professional skill.
Development of science-based analytic technique has been essential to
the reliability and credibility of EIA. Yet the professionalization of EIA
entails a predictable risk – the adumbration of purpose by technique.
(Caldwell, 1989: 8)

The origins for this failure can be traced to the way in which agencies have
chosen to implement impact assessment provisions. For example, it is clear
that the intent of NEPA 'was not a policing action, but rather a device to
transform and reorient the values and assumptions' of Federal agency
decision making (Caldwell, 1988a: 77). However, successive administrations
in the United States 'have chosen to leave the interpretation and
enforcement of NEPA largely to the courts. This has resulted in an emphasis
on the judicially enforceable aspects of NEPA to the neglect of its
substantive provisions' (Caldwell, 1988a: 78). Thus, even though the science
of impact assessment still requires further improvement, the basic problems
with impact assessment are 'less scientific and technical than political'
(Caldwell, 1988a: 80). It is only through reconceptualization, and a
consideration of process as well as methodology, that impact assessment will
achieve its full promise.

There is an expectation that impact assessment will eventually be an
integral component of the normal processes of planning and decision making
rather than just a check upon them. It is a transformation that 'requires
some fundamental rethinking of conventional assumptions about bureaucratic organizations and control' (Caldwell, 1988a: 82). If impact assessment
is to evolve beyond its present narrow methodological confines, attention
must be paid to the institutional arrangements that define the role of impact
assessment in resource management.

CHAPTER 3

Institutional arrangements for impact assessment

Introduction

Institutional arrangements define the conditions under which resources are managed. They affect the implementation of resource policies and they structure the policy-making process (Smith, 1984b: 1). Institutional arrangements are a composite of legal powers, administrative structures and financial provisions, which give rise to a definable system of public decision making. It is these institutional arrangements that define the role played by impact assessment in resource management and environmental planning.

This chapter outlines the prerequisites for effective institutional arrangements. It begins by examining the central role of the law. Legal systems vary greatly from nation to nation but there are certain inherent constraints within most systems of environmental law that limit the range of institutional arrangements for the implementation of impact assessment.

Institutional arrangements involve not only administrative structures but also the decision rules, mandates, finances and operations that affect resource management. How then are institutional arrangements to be defined and operationalized in the study of impact assessment? Various definitions for institutional arrangements are outlined, guidelines for institutional policy analysis are examined, and a framework for the analysis of institutional arrangements is detailed. The chapter then reviews the empirical work on institutional analysis in impact assessment, with the discussion focusing on the dominant theme that has emerged thus far: the consideration of impact assessment as a policy strategy.

The role of environmental law

Legal provisions for environmental protection, planning and regulation establish the context for impact assessment. These provisions vary from country to country and are a product of each nation's distinct political culture. Indeed, as much as the law may be viewed as an instigator of reform, it is itself shaped by new political perspectives (O'Riordan, 1981: 265). Thus, it is not surprising to find that Western nations have evolved systems of law that 'favour use over preservation, private property rights over common property rights, and the generation of wealth and productivity over amenity' (O'Riordan, 1981: 265).

In broad terms, Western nations have two systems of law that can apply in environmental situations: common law and statutory law (O'Riordan, 1981). Common law is the law of precedent, established by judicial rulings and subject to interpretation by judicial proceedings. Statutory or public law is the law of the constitution and of governments, established by legislative assemblies (national, regional and/or local) and subject to revision and change through political processes.

Common law provides for two types of remedy: compensation for damages and the imposition of an injunction, or court order, suspending the offensive activity. These may be granted on the basis of the common law theories of:

- *Nuisance*: the concept of nuisance is based on a basic rule that property owners should use property in a manner that is not injurious to that of another. Where this mutual obligation between landowners has been breached, nuisance theory provides for damages or an injunction provided the plaintiff can show the damage to be the fault of the defendant, that 'substantial' injury can be demonstrated and that the injury is 'unreasonable'.

- *Trespass*: this is said to have occurred when there has been a physical invasion of private property by persons or polluting material, constituting an invasion of the right to exclusive possession of the land.

- *Negligence*: the basis of negligence is carelessness. The law assumes that property owners are totally responsible for the consequences of their actions. Thus, a defendant can be sued if he or she has failed to take due precautions or adhere to an accepted standard of practice.

- *Liability*: strict liability pertains to activities that endanger the public and for which the defendant can be held to be socially responsible.

- *Riparian rights*: these exist where a property owner's land or interests touch a watercourse or body, such as a river or lake. Riparian rights are based on the principle that upstream uses of the water resource should not impair the rights of downstream property owners to use that same resource. Hence, they are normally used to settle issues of water supply, allocation and pollution.

In practice, the extensive use of common law for environmental protection has been widely precluded by the rules of precedence and an absence of clear rulings in environmental cases. In addition, aspects of common law theories inhibit their use in environmental situations. For example, strict liability requires that a precise cause of damage and lack of prevention must

31

be shown. Meanwhile, widespread use of nuisance and trespass has been restricted by problems associated with the requirements of standing, whereby litigable rights apply only to an individual who owns property or can demonstrate significant economic damages as a consequence of offending activities. Moreover, unsatisfactory findings, such as the award of damages rather than the imposition of an injunction and the cessation of activities, have often resulted from legal actions under the provisions of common law (O'Riordan, 1981; Wenner, 1989, 1990).

As a reaction to these restrictions on the use of common law to redress damages, Sax (1970) proposed that the doctrine of public trust be employed to ensure the protection of the environment. Public trust doctrine stipulates that governments have a legal responsibility to hold certain common property resources such as air, water and mountains in trust for the general free and unimpeded use of the public. The logical extension of this doctrine is the concept of citizen environmental rights, first codified by Sax when he drafted the 1970 Michigan Environmental Protection Act (MEPA).

The MEPA was one of the first statutes to provide for environmental protection, and since the late 1960s and early 1970s, numerous other statutes have been passed in the three main areas of environmental law: pollution control, resource conservation and land use control. Because it emphasizes prevention rather than remediation, statutory law has become the predominant element of contemporary environmental law. Indeed, with minor exceptions, 'the field of environmental law is entirely statutory' (Grad, 1985: 13). Much of this environmental legislation has attempted to redress the limitations of common law by relaxing the strict requirements of standing and providing opportunities for public participation in environmental decision making (Wenner, 1990).

However, there are several persistent problems with statutory environmental law, including the disparity in resources for litigation, a lack of effective sanctions or the failure to enforce sanctions, and the apparent inability of law to accommodate issues of technical and scientific uncertainty (Grad, 1985). In addition, statutory law has often moved the 'burden of resolving uncertainty to the courts', which has cast judges in an entirely new role as 'quasi-legislators and quasi-administrators' in implementing provisions for environmental protection (Wenner, 1990, 191). Particularly in the United States, this new role for the courts has been the source of considerable controversy and debate (see Wenner, 1989, 1990).

These problems are recurring and represent very real constraints on the use of environmental law. Moreover, they are intrinsic to the use of law for environmental planning. For example, McAuslan (1980) has suggested that the law relating to land use planning is, by definition, neither objective nor neutral. Furthermore, 'the law, its administration and official interpretation . . . is itself a major contributory factor to the continuing disarray of planning' (McAuslan, 1980: 2). McAuslan (1980) identified three competing ideologies for the role of planning law:

- *Private property ideology*: based on the traditional common law approach to the role of law. This suggests that the law exists and should be used to protect private property and its institutions.
- *Public interest ideology*: the role of law as seen by the public administrator. The law provides the backing and legitimacy for planning and programmes that advance the public interest, often and if necessary against the interest of private property.
- *Populist ideology*: the law is a vehicle that should be used to promote democracy and justice through the advancement of public participation. This approach is counter to both the orthodox public adminstration approach to the public interest and the common law approach to the overriding importance of private property.

These three ideologies were used by McAuslan to examine planning law and its administration. The central premise of his book is that the major environmental challenges facing society involve not only substantive issues of environment and resource management but, more importantly, *procedural issues* which he defined as 'the style and nature of government; the structure of decision-making and the laws governing the decision-makers and establishing the programmes of administration' (McAuslan,1980: xi).

As one of the notable products of societal attempts to deal with environmental issues, impact assessment represents a major challenge to the 'procedural issues' of environmental and administrative law. Law provides a basis for planning and regulation but it must be viewed, understood, administered and implemented within a wider political context. That is, the implementation of provisions for impact assessment necessitates attention to the institutional arrangements for management.

Institutional arrangements defined

The wide range of definitions applied to institutional arrangements have been well reviewed by Mitchell (1989: 242–5). The most complete definition is that of Craine (1971: 522) who indicated that the study of institutional arrangements requires:

> special attention to the configuration of relationships
> (1) established by law between individuals and government;
> (2) involved in economic transactions among individuals and groups;
> (3) developed to articulate legal, financial and administrative relations among public agencies; and
> (4) motivated by social–psychological stimuli among groups and individuals.

Craine's definition also requires that the emphasis in the study of institutional arrangements be upon the linkages among the variables as they pertain to authority and action within a public decision making system.

Other definitions of institutional arrangements have referred to a 'cluster' of customs, norms and laws; 'forms' of governance designed to influence human behaviour; and 'the pattern of agencies, laws, participants and policies' that structure resource management (Mitchell, 1989: 244). Noting that institutional changes often involve fundamental shifts in income, power, and prestige, Ingram et al (1984: 323) stated that the 'term "institutional" is meant to include those legal, political and administrative structures and processes through which decisions are made with respect to public policy'. The range of variables identified by Ingram et al (1984) as affecting institutional arrangements included laws, regulations, informal procedures and the distribution of political support. They also suggested that institutional arrangements are affected by public opinion, the attitudes and preferences of interest groups and the orientation of public officials (Ingram et al, 1984).

Several terms have been used to identify the people and interests seeking representation within institutional arrangements for management. For example, Mitchell (1989) referred to key participants or actors. Actors in resource management have been best defined by Royston and Perkowski (1975: 137–8) as 'those who claim a priority interest in a particular resource' and are active in the defence of that resource. Here, the term *stakeholders* is used, as it is more inclusive of the full range of participants in a decision environment. Stakeholders include both the *effecting interests* (resource developers, implementers and regulators) and the *affected interests* (the beneficiaries, victims, those interested in the plan) capable of exerting influence on the outcome of decision making (Lang, 1986b: 27–8).

These various aspects may be summarized by stating that institutional arrangements refer to 'a definable system that provides both opportunities for and constraints upon policy making' (Mitchell, 1987: 7). The key aspect is the focus on the interaction of several variables (Mitchell, 1989; Smith, 1984b), most importantly:

- legislation and regulations
- policies, guidelines and activities
- administrative structures
- economic and financial arrangements
- political structures and decision processes
- historical and traditional customs and values
- stakeholders.

Guidelines for institutional policy analysis

In his review of the field, Gormley (1987) suggested that while the practice of institutional policy analysis is extensive, understanding remains poor: 'those who practise institutional policy analysis seldom think of themselves as institutional policy analysts. Rather, they think of themselves as experts on legislative, executive, judicial or electoral behaviour. As a result, the institutional policy analysis literature, though rich, remains scattered and diffuse' (Gormley, 1987: 153). Institutional policy analysis is the study of government reform and its consequences. It emphasizes the redesign of governance and attention to issues of political architecture, such as: procedural choices, the redefinition of relationships within government, and decisions that affect the influence of outsiders on government itself (Gormley, 1987). At its best, institutional policy analysis is the application of clear values and systematic empirical research to a problem of institutional design.

Institutional policy analysis tends to emphasize procedural elements and values. There are several reasons for this orientation (Gormley, 1987). First, procedural effects are often perceived to be more easily measurable than actual behavioural impacts. Second, as issues become more complex and factual evidence more uncertain, policy outcomes become harder to predict. Therefore, stakeholders must agree to abide by outcomes of administrative processes in lieu of any substantive certainty. Third, procedural concerns are more consistent with Western political culture than substantive concerns. An emphasis upon procedure need not be a weakness, but it is vital that substantive criteria be applied to problems of institutional design. Too often, however, this is not done and Gormley (1987: 157) is justly critical of the literature on institutional policy analysis for its reliance upon a single method to examine a single reform of a single institution.

To counter these failings, Gormley (1987) advocated a number of remedies. Analysts should consider both formal and informal reforms in the design of solutions to institutional problems. Furthermore, the reliance upon *coercive* controls should be balanced by the adoption of *catalytic* controls. Typically, attempts to control the policy process have utilized coercive means such as laws with a predetermined time-frame for implementation, programme audits, sanctions and legislative vetoes. Catalytic controls differ in that they provide an impetus for change but do not stipulate how that change should occur. For example, mechanisms such as public utility commissions, consumer advocacy offices and 'agenda-forcing' statutes require action and provide policy direction, but do not limit the capacity for creative problem solving within the policy process (Gormley, 1987: 160). Typically, impact assessment provisions are an example of catalytic control within environmental policy.

Policy analysis criteria need to be much more clearly and precisely defined (Gormley, 1987). For example, frequently used criteria include terms such as responsiveness, accountability, leadership, effectiveness, fairness and equilibrium. Rarely are these terms defined clearly or in a fashion that clarifies their measurement. More precise definitions of policy analysis criteria are essential if these criteria are to promote communication and not obfuscation. Lastly, institutional policy analysis would be improved if analysts used more than one method, compared reforms and examined more than one political institution (Gormley, 1987).

Similar guidelines for improving the standard of institutional policy analysis have been proposed by Ingram et al (1984). They urged that analysts should pay particular attention to three main aspects of institutional arrangements:

- actors and their stakes in decision making
- the resources actors have at their disposal,
- the biases of the alternative decision making arenas through which actors may try to reach their goals.

The analysis of group interests, resources, membership and practices is a necessity in understanding institutional arrangements. The analyst must identify anticipated and perceived, as well as scientifically predictable, impacts. Thus, the definition of interests in institutional arrangements should be made on an understanding of perceived concerns rather than administrative or political boundaries. It is important to remember that changes in the issue context will affect perceptions of interests, the emergence of participants and the definition of relevant stakeholders.

All stakeholders have a range of resources at their disposal. These include such elements of the policy process as legal rules and arrangements, economic power, prevailing values and public opinion, technical expertise, control of information, the control of organizational and administrative mechanisms, and political resources. These resources are rarely even in their distribution. Moreover, the imbalance of resources between stakeholders may be further exacerbated by inherent biases in the forums available for decision making such as legislatures, courts, administrative agencies, regulatory tribunals and public hearings. Thus, it is important that the analyst remembers that while most decisions involving resources are made 'on the basis of bargaining, negotiation and compromise . . . the character of the bargaining varies, depending on the arena in which the bargaining occurs' (Ingram et al, 1984: 328).

To overcome the impediments to effective institutional arrangements, Ingram et al (1984) recommended several different strategies and techniques. Their range of solutions included modifications to existing arrangements (e.g. by changing management practices or through changes in legal definitions, rights and relations), the creation of new arrangements (e.g. new institutions and/or engineering and technical advances) and the use of different approaches to resolve allocative conflicts (e.g. the operation of market mechanisms, the use of negotiated settlements or the dissemination of newly developed information).

The solutions proposed by Ingram et al (1984) focus on the need to improve the capacity for institutional arrangements to provide for integration in the management of resources. This is the same goal as Mitchell and Pigram (1989) who assessed the application of integrated resource management in the Hunter Valley, New South Wales. According to Mitchell and Pigram (1989: 198) the 'purpose of an institutional analysis is to identify and assess the leverage points at which it is possible to improve resource management'. To provide for this analysis, Mitchell and Pigram (1989) utilized a framework originally devised by Mitchell in 1987 and updated in 1990 (Mitchell, 1987, 1990a).

The Mitchell framework focuses upon the provision of institutional arrangements that provide for co-ordination and an integrated approach to resource management. The framework assumes that the responsibilities for public authority in resource management are fragmented and shared among several different agencies. This gives rise to boundary problems among and within management agencies, generating problems of overlapping jurisdictions and multiple mandates. Institutional arrangements must counter these problems if they are to facilitate integrated resource management. However, in developing institutional arrangements it is important to 'recognize that numerous options are available, all with strengths and weaknesses. Since every option will be imperfect, attention should be directed towards identifying the mix of leverage points which can be used collectively to achieve co-ordination' (Mitchell, 1990a: 17).

The framework has six components, each representing a *leverage point* where opportunities exist to improve integration:

- *Context*: the opportunities and constraints provided by broader contextual aspects relating to the state of the environment, prevailing ideologies, economic conditions and the existing pattern of legal, administrative and financial arrangements. With respect to impact assessment, factors that could influence the context would include such factors as attitudes towards regulation, the tradition of environmental law and support for environmental protection.
- *Legitimation*: the presence of statutory powers, political commitment and administrative policies that identify agency objectives, determine the responsibility, power or authority of agencies, clarify rules for intervention and the resolution of boundary problems. For example, are impact assessment provisions enshrined in law, and how are those provisions being implemented?
- *Functions*: the consideration of which management functions are assigned and at what scale. These should be linked explicitly to legitimation and to the structures. It is important for studies of institutional arrangements to indicate which resource management functions are being appraised, as different functions may require different institutional arrangements. These should include both generic functions and substantive functions. Generic functions in resource management include activities such as surveillance (the collection and reporting of information), mediation (the development of joint strategies and the resolution of conflicts) and control (the application of regulatory responsibilities and implementing authority). In impact assessment, typical substantive functions include better decision

making, environmental planning, impact mitigation and conflict resolution.

- *Structures*: the provision of appropriate organizational structures. These should be related to the desired functions and considered on the basis of efficiency, accountability and flexibility. For example, should impact assessment procedures be administered by a central agency or an existing department with an environmental mandate?
- *Processes and mechanisms*: the facilitation of bargaining, negotiation and mediation through political and bureaucratic processes, informal mechanisms and provisions for wider input (such as regional planning, benefit–cost analysis, impact assessment or public participation). What is the role of impact assessment *vis-à-vis* these other mechanisms, and how are interests represented within impact assessment processes?
- *Organizational culture and participant attitudes*: the identification of characteristics of the organizational culture that might provide either incentives or disincentives for an integrated approach to resource management. For example, attitudes regarding sustainability might affect the implementation of impact assessment.

A modified version of this framework is presented in Fig. 3.1. Mitchell's original framework has been amended to incorporate the various suggestions forwarded by Gormley (1987) and Ingram et al (1984). Figure 3.1 also integrates more fully the key aspects defining institutional arrangements.

The amended framework presents a series of leverage points at which institutional arrangements may be influenced, modified and/or revised. The leverage points are not related in a hierarchical manner but, rather, in an iterative fashion wherein changes to any one component are likely to affect other aspects of the system. The framework may be used as a *normative model* to guide the design of institutional arrangements, their modification and/or revision. It can also be used as a *descriptive schema* for institutional policy analysis and the evaluation of institutional arrangements for management. The framework is used in both capacities within this text. In Chapter 6, the framework is used normatively when it is integrated into an overall model redefining impact assessment. In the subsequent case study chapters (7–9) the framework is used in the evaluation of the international experience with impact assessment in a number of areas.

Institutional analysis and impact assessment

There is growing recognition of institutional arrangements in empirical work on impact assessment. This has been shown by the publication of theme issues of journals on policy and institutional aspects of impact assessment (e.g. the 1988 special issue of *Impact Assessment Bulletin* edited by Bartlett) and an increasing number of journal articles focusing upon changing impact assessment provisions in Europe (Wood and Lee, 1988; Wood and McDonic, 1989), New Zealand (Morgan, 1988), India (Vizayakumar and

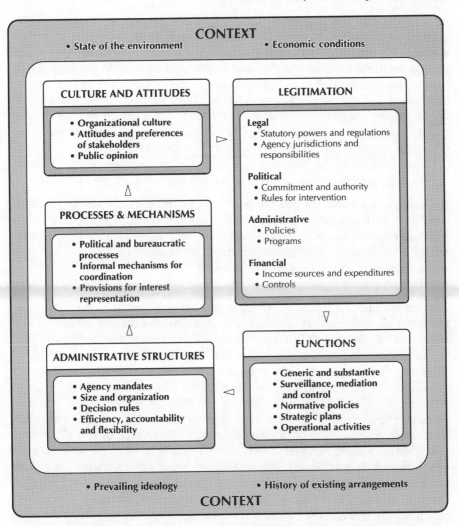

Fig. 3.1 A framework for the assessment of institutional arrangements
     Source modified from Mitchell (1990a)

Mohapatra, 1991), the Philippines (Abracosa and Ortolano, 1987), Canada (Smith, 1990b) and Indonesia (Stern, 1991). Attention to cross-national comparisons of impact assessment provisions (O'Riordan and Sewell, 1981; Wathern 1988b) and a focus on institutional questions at international conferences (Paschen, 1989) are also indicative of this trend. As yet, however, formal research on institutional arrangements for impact assessment is still at a formative stage and the field lacks any definitive or seminal works.

One country where research advances have been made is Canada. Because of the division of powers and responsibilities between the federal government and the provinces, the issue of institutional arrangements has been of particular importance in Canada. Research on the status of impact assessment provisions in Canada has included: evaluations of changes to the federal environmental assessment and review process or EARP (Rees, 1980; Fenge and Smith, 1986), a comprehensive review of the status of impact assessment in Ontario (Gibson and Savan, 1986; Gibson, 1990), an analysis of the impact and effectiveness of institutional changes to the system for impact assessment in British Columbia (Smith, 1988a), and overviews of the Canadian scene (Smith, 1987b, 1989, 1990b; Smith et al, 1989).

In his analysis of the Ontario legislation for impact assessment, Gibson (1990) suggested a set of principles for effective impact assessment:

- The impact assessment process must be enshrined in law and compliance with its procedures should be legally enforceable.
- There must be comprehensive, early and clear application of the process.
- The consideration of alternatives must be mandatory.
- 'Environment' must be broadly defined.
- Effective public participation is essential.

His findings indicated that 'the Ontario process stands as an imperfect model for other jurisdictions' (Gibson, 1990: 79). This suggestion reinforces the viewpoint of Smith (1990b) that neither Ontario nor the federal EARP represents a good normative model for impact assessment in Canada.

Smith (1989) conducted a comparative evaluation of impact assessment provisions for each jurisdiction in Canada (the 10 provinces and the Federal government: Fig. 3.2). The evaluation revealed three categories of impact assessment provision. All jurisdictions were found to have a good awareness of the role and function of impact assessment. The variability arose in how that function should be carried out, on what projects and with what level of commitment. It was found that a lack of enabling legislation and poor institutional arrangements had restricted impact assessment in four jurisdictions. In addition, a strong definition of 'environment' and good procedures for public participation were found to be key criteria differentiating provisions (Smith, 1989).

Using these data, Smith (1990b) indicated that the provisions for impact assessment in Saskatchewan and in Newfoundland were a much better normative model for the Canadian situation than the more well-known and frequently cited examples of Ontario and the federal EARP. However, this research highlights the difficulties involved with conducting comparative evaluations in institutional policy analysis. As noted previously, Gormley (1987) had advocated that policy analysts compare reforms and examine more than one political institution. Smith's evaluation compared provisions for impact assessment for 11 jurisdictions with the constraint that changes and modifications are an ongoing and constant feature of the policy environment. Thus, the research findings were transitory and any conclusions had to be tempered by a caveat that these were a 'passing snapshot' of

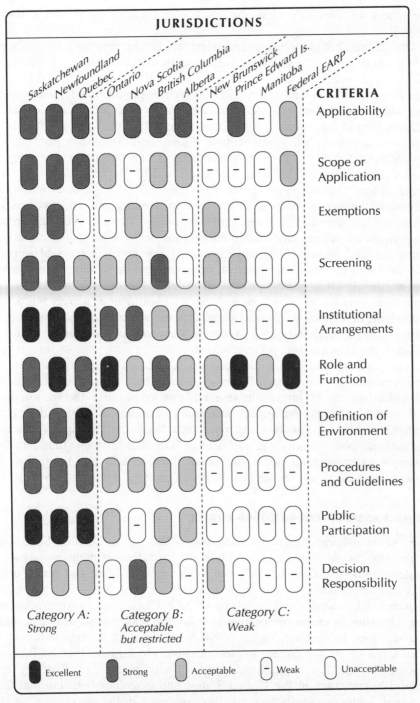

Fig. 3.2 An evaluation of Canada's impact assessment provisions
Source after Smith (1989, 1990b)

the status of Canada's changing impact assessment provisions (Smith, 1990b).

The Canadian situation appears to exemplify the prognosis of Caldwell (1989: 12) that impact assessment 'will be most effective where environmental values (1) are implicit and consensual in the national culture and (2) are explicit in public law and policy'. In Caldwell's opinion, impact assessment has had a significant but imperfect influence upon policy and decision making. Impact assessment must be more than a technical process. It should represent a strategy for administrative reform and establish a principle of environmental policy central to governance (Caldwell, 1989).

This line of thinking has also been expressed by Bartlett (1989b: 3). He wrote: 'More than methodology or substantive focus, what determines the success of impact assessment is the appropriateness and effectiveness in particular circumstances of its implicit policy strategy.' In the past, proponents of impact assessment have often been rather naive about its policy effects. Mostly, they have assumed the policy effects of impact assessment to be inherently positive, often misreading or misunderstanding the nature of the policy process (Bartlett, 1989b; Taylor, 1984: 7). Conversely, critics often have viewed impact assessment as merely symbolic, redundant, diversionary, obstructionist, expensive, wasteful and/or unnecessary. Again, there has been a failure to view impact assessment within a broader institutional and policy context.

These failings underscore the need for further attention to institutional arrangements in the area of impact assessment. This need has been recognized in the literature but research that emphasizes the institutional and policy contexts for impact assessment remains a new and largely undeveloped field. The one exception has been the principal focus for institutional policy analysis in the field of impact assessment thus far: the attempts to understand better the possibilities, effects and limitations of impact assessment as 'an adjustable *policy strategy*' (Bartlett, 1989b: 3).

Impact assessment as policy strategy

The National Environmental Policy Act (NEPA) in the United States has been used as an illustrative example of impact assessment as a policy strategy by both Caldwell (1982) and Taylor (1984). Their two books offer different approaches to the subject matter. Caldwell (1982) is a more conventional summary and discussion of NEPA as a strategy for procedural reform, using an inside perspective to detail the background story to events and activities. In contrast, Taylor (1984) focused not on the establishment of NEPA but on its subsequent implementation. He suggests that NEPA was responsible for the institutionalization of environmental values by the federal bureaucracy in the United States, and examines both internal and external factors contributing to this success. Both works are worth examining in further detail.

Caldwell (1982) describes the use of NEPA to invoke a procedure for environmental impact analysis and to effect a major reorientation of public policy and administration. The role of the environmental impact statement (EIS) in this procedure was 'conceived as a mandatory, action-forcing reorientation of planning and decision making' (Caldwell, 1982, 1).

From Caldwell's perspective, the EIS was conceived to provide an effective, policy-focused use of scientific knowledge and methods as the means to achieve environmental impact analysis. In his view, 'NEPA is exceptionally coherent. It declares a policy which it illustrates by example, specifies procedural means toward achievement of this policy, and provides institutional arrangements and jurisdictional safeguards to ensure that these procedures and the objectives they serve will not be lost in the complex interactions between politics and bureaucracy' (Caldwell, 1982: 11).

Thus NEPA required a reconsideration of the use of science in decision making. It also necessitated several administrative adjustments by federal agencies (Caldwell, 1982: 58–9):

• organizational integration of EIS functions into existing structures and procedures
• the reform of administrative procedures
• the hiring of new personnel, especially the required impact assessment analysts
• added budgetary requirements
• added public consultation responsibilities.

Adapting McLuhan's aphorism that the 'medium is the message' to read 'the procedure is the policy', Caldwell suggests that NEPA has been influential in fostering the use of science within public policy and administration, and in promoting public participation in environmental issues. Mostly, however, NEPA was enacted 'to achieve or restore a balance in American public policy that many persons believed had been lost in an overemphasis on economic, technological, and development values' (Caldwell, 1982: 139–40).

Caldwell presents a detailed summary of NEPA and its intent as a strategy for procedural reform. However, his commentary appears somewhat optimistic regarding NEPA's implementation. In practice, the EIS mechanism has not always functioned as envisioned. Rather, it has often become an end in itself and the focus of protracted conflicts. Moreover, as outlined in Chapter 2, the adherence to a philosophy of impact assessment through impact statements has had a nullifying effect on the evolution of impact assessment practice. Rather than acting to broaden the basis of environmental planning, the EIS has often served a reductionist role, limiting decision making to those factors that are easily measured, understood or defended, particularly in adversarial situations. This is not what was intended, nor does it accord with the model of NEPA detailed by Caldwell (1982).

Taylor (1984) addresses problems with uncertainty and the ability of impact assessment to provide a process of 'science' within administrative decision making. He suggests that the two are not reconcilable. Science seeks truth and absolutes, while administrative decision making is a process of political bargaining with neither fixed limits nor pretence of objectivity. Thus, expecting impact assessment to provide both scientific rigour and efficient decision making is akin to putting a square peg into a round hole.

Taylor's approach was to compare the impact assessment process under NEPA in two different agencies, the Forest Service and the Army Corps of Engineers, over an extended period. Variants of the impact statement system in other policy areas were also reviewed. Three questions were posed (Taylor, 1984: 5):

- To what extent (and how) did NEPA succeed in institutionalizing a greater sensitivity to environmental concerns in the federal bureaucracy?
- How widely applicable was NEPA's impact statement mechanism as a strategy for enhancing the reasoned consideration of issues by public organizations?
- Under what conditions can science-like rules and procedures actually help rather than hinder democratic politics?

His central premise was that NEPA's use of the EIS process was 'an attempt to import "scientific" norms and procedures into a political setting of intense conflict' (Taylor, 1984: 8). Taylor then added the corollary that the ability of an organization to solve problems is a function of its willingness and capabilities to adopt a strategy for learning: 'the series of decisions that an organization makes over time is more important than any particular decision, and the growth in knowledge relevant to the organization's work is critical' (Taylor, 1984, 22).

Taylor outlined the basis for scientific understanding and compared this model to the statutory framework for environmental analysis established under NEPA. In his view, a system of formal analysis must:

- focus on important issues
- specify how much detail must be provided for various kinds of analysis
- prevent the manipulation of alternatives to obscure the real choices available
- facilitate helpful criticism by informed outsiders
- provide forums for resolving technical disputes
- adjust the burden-of-proof rules or distribution of analytical resources to make the system workable if the resources of outsiders and insiders are greatly out of balance
- provide incentives for the analysis actually to be used in decision making
- encourage continual improvement of analytical methodology.

Regarding the implementation of impact assessment under NEPA, Taylor's results suggest that when 'inside analysts' were able to explore possible environmental trade-offs 'environmentally better decisions' were likely to result (Taylor, 1984: 251). In addition, he found that all projects benefited from relatively inexpensive environmental mitigation. However, Taylor (1984: 251) also cautioned that 'environmental values remain "precarious" values: though they are now formally a concern of every agency, the natural bias of the organization will go against their realization unless they are protected by special arrangements.' More generally, Taylor found that impact assessment offered an effective alternative to science and the 'partisan mutual adjustment' approach of politics as a means of balancing societal trade-offs in decision making (Taylor, 1984: 302–7).

Taylor's book represents a considerable breakthrough in impact assessment research. Not only does he focus on impact assessment as a policy strategy but he seeks to place impact assessment in a wider context by adopting as his frame of reference the roles of science and politics in societal decision making. Taylor's work has already stimulated other studies in the United States. For example, Bartlett and Baber (1989) have examined whether impact assessment can enhance bureaucratic expertise in dealing with complex and interdependent policy issues and, thus, improve government effectiveness. They assessed the response of impact assessment (NEPA) to three central challenges of public administration:

- *effectiveness*: as determined by programme evaluations and designed to improve the future performance of administrative decision making
- *co-ordination*: either through authoritative coercion and central co-ordination or through bargaining and mutual adjustment
- *legitimacy*: the issues of bureaucratic accountability and the role of public participation.

Consistent with the findings of Taylor, Bartlett and Baber found that impact assessment can improve the scientific content of agency decision making and enhance the effectiveness of bureaucratic actions. Moreover, while the evidence for greater regional co-ordination as a result of NEPA was 'mixed', it was concluded that impact assessment does force 'agencies to engage in a greater sharing of the scientific and technical information upon which their claims to expertise are founded' (Bartlett and Baber, 1989: 151).

Bartlett and Baber also found that NEPA had fostered an enhanced sense of agency legitimacy regarding environmental decision making. Impact assessment provided the means for agencies to resist outside pressures to whatever extent they were 'able and motivated to do so', with the corollary that the already rich and powerful agencies become more so as a consequence of using the opportunities afforded them by impact assessment (Bartlett and Baber, 1989: 152).

Wandesforde-Smith and Kerbavaz (1988: 162) have suggested that 'the tendency has been to see the essential dynamic in the history of EIA in the United States as arising out of a tension and conflict between recalcitrant bureaucratic insiders and reform-minded external intervenors.' They credit Taylor with breaking this impasse and moving research towards an identification of the conditions that sustain or erode the ability of stakeholders to contribute to impact assessment processes. Wandesforde-Smith and Kerbavaz (1988) have attempted to move beyond the findings of Taylor. Public policy instruments differ in their opportunities for bargaining and input into the making and implementation of policy. An important consideration, therefore, in understanding the way different instruments work in practice rests on the determination of 'how a combination of resources, political circumstances, and skill relating the two empowers some to bargain more effectively than others, and thus to shape the evolution of policy' (Wandesforde-Smith and Kerbavaz, 1988: 161).

Wandesforde-Smith and Kerbavaz (1988) defined this combination as the product of political entrepreneurship. They suggest that it has been the principal factor in the evolution of impact assessment in the United States and California, and argue that it is critical that individuals and their roles be recognized and included in the evaluation of policy change. In a subsequent paper, Wandesforde-Smith (1989: 161) has suggested that impact assessment 'survives and prospers . . . as a policy instrument not because structural variables determine that result but because EIA affords such rich and diverse opportunities for political entrepreneurs to alter the circumstances of their existence.'

Administrative entrepreneurs use the impact assessment process to build coalitions to aid them in achieving change, make use of opportunities for strategic analysis to enhance the technical expertise of agencies and affirm and develop environmental values. Thus, in an era of environmental uncertainty and rapidly changing goals and objectives, it is administrative entrepreneurs that 'alter the structure of constraints on the exercise of choice' (Wandesforde-Smith, 1989: 162).

This work implies that an explicit behavioural element must be incorporated to complement the legal and administrative aspects tradition-ally included within institutional policy analysis. Moreover, if impact assessment is to be understood as a policy strategy, then a framework for the evaluation of policy implementation is needed that encompasses:

- the structural elements of reform analysed by both Caldwell and Taylor
- the behavioural elements emphasized by Wandesforde-Smith
- the macro-level elements identified by the literature on institutional arrangements.

Sabatier and Mazmanian (1981) have developed an analytical framework to evaluate policy implementation that appears to accommodate these necessary components (Fig. 3.3).

The central premise of Sabatier and Mazmanian's framework is the need for both conceptual development and empirical examination of the linkage between individual behaviour in the implementation of policy and the political, economic and legal context of the policy environment (Sabatier and Mazmanian, 1981: 5). Sabatier and Mazmanian (1981: 3) perceived the policy implementation literature to be heavily reliant upon the specific and

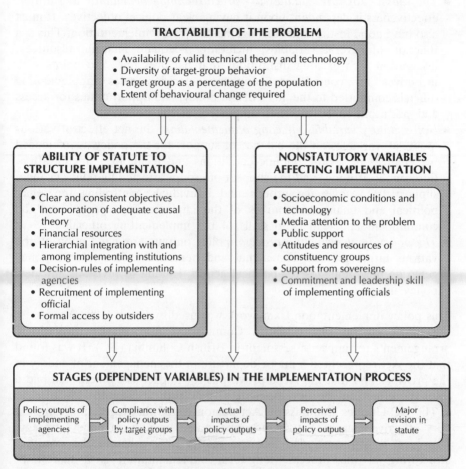

TRACTABILITY OF THE PROBLEM

- Availability of valid technical theory and technology
- Diversity of target-group behavior
- Target group as a percentage of the population
- Extent of behavioural change required

ABILITY OF STATUTE TO STRUCTURE IMPLEMENTATION

- Clear and consistent objectives
- Incorporation of adequate causal theory
- Financial resources
- Hierarchial integration with and among implementing institutions
- Decision-rules of implementing agencies
- Recruitment of implementing official
- Formal access by outsiders

NONSTATUTORY VARIABLES AFFECTING IMPLEMENTATION

- Socioeconomic conditions and technology
- Media attention to the problem
- Public support
- Attitudes and resources of constituency groups
- Support from sovereigns
- Commitment and leadership skill of implementing officials

STAGES (DEPENDENT VARIABLES) IN THE IMPLEMENTATION PROCESS

| Policy outputs of implementing agencies | Compliance with policy outputs by target groups | Actual impacts of policy outputs | Perceived impacts of policy outputs | Major revision in statute |

Fig. 3.3 The policy implementation framework of Sabatier and Mazmanian (1981)

unique details of individual case studies and programmes, ignoring the broader legal and political contextual variables that structure the policy process. They were also critical of existing analytical frameworks for seriously underestimating the ability of a statute to *structure* the policy implementation process (Sabatier and Mazmanian, 1981: 5).

To counter these weaknesses, the Sabatier and Mazmanian framework (Fig. 3.3) explicitly recognizes the potential for a statute to structure the policy implementation process. Within the framework, public policy is viewed within four broad categories:

- *The tractability of the problem*: some problems are easier to solve than others. Problem solutions are influenced by uncertainty; the extent and diversity of behavioural change desired; and, the nature and characteristics of the population that is the 'target' of policy change.

- *The ability of the statute to structure implementation*: a statute's effectiveness is dependent upon it having clear, logical objectives. It must also have good institutional arrangements for its implementation. This is a function of several variables, including: adequate financial resources; integration with existing institutions that balances few veto points with incentives for compliance; supportive decision rules; implementing officials committed to the goals of the policy; and opportunities for access and participation by external supporters of the policy.
- *Non-statutory variables affecting implementation*: the net effect of various 'political' variables on the balance of support for the policy as influenced by: changes in socio-economic conditions and technology; media attention; public attitudes and the support of groups affected by policy change; support from sovereign groups and individuals controlling the legal, political and financial resources of the implementing agency; and the commitment and leadership skills of the implementing officials.
- *The policy implementation process*: policy outputs are acted upon by the various target groups. The actual and perceived effects of the policy generate new demands for political action, leading to revisions and new policy initiatives.

This policy implementation framework was used by Smith (1988a) to assess the effect of the 1980 Utilities Commission Act on the institutional arrangements for impact assessment in British Columbia (BC). It was found that the Act represented a major departure in the management of energy in the province and that the Act's provisions for impact assessment had had a dramatic effect on the development of new projects (Smith, 1988a).

The BC Utilities Commission Act demonstrated that legislation can be a very effective means of invoking impact assessment as a policy strategy. However, it was concluded that 'no matter how coherent the statutory structuring of the process, long-term implementation is a matter of continued political commitment and action' (Smith, 1988a: 442). Support for this view also is evident in the work of Wathern (1988c; 1989) concerning the implementation of the European Economic Community (EEC) Directive on EIA in Britain. Wathern has demonstrated that the response to the EEC Directive in Britain has been an attempt to 'contain the impact of a directive which was seen to carry significant administrative costs, but few political benefits' (Wathern, 1989: 36). The inference to be drawn from both studies is that the long-term effectiveness of impact assessment mostly depends upon the political variables of the public policy process.

Summary

Institutional arrangements can provide provisions for impact assessment that enable an effective policy strategy to be implemented. The factors affecting the effectiveness of those provisions offer a range of leverage points from which the provisions may be modified, amended or curtailed. Within this framework, the success of impact assessment as a policy strategy appears to be dependent upon:

- the strength of the legislative base
- the intended role and function of impact assessment
- political commitment
- the roles played by three key types of stakeholder:
  - · the internal analyst-advocate
  - · the administrative entrepreneur
  - · external pressure groups.

Research focusing upon institutional arrangements for impact assessment remains at a formative stage. There is a need for more comparative work. Moreover, evaluations must begin using formal frameworks for analysis if the field is to progress beyond a simple description of prevailing legislation, administrative structures and interested actors. Thus far, the level of conceptualization has served to focus attention on the identification of the conditions that sustain or erode the ability of stakeholders to contribute to impact assessment processes. In turn, this underscores the need for a clear understanding of the policy process and the roles of stakeholders in the formation of public policy.

CHAPTER 4

Public policy and interest representation

Introduction

Political commitment is a critical determinant of impact assessment effectiveness. The factors that influence political commitment, and the mechanisms by which that influence can be exerted within the public policy process, constitute the focus of this chapter.

In its simplest terms, the study of public policy is the study of what governments do. Public policy may be stated, implied, perceived or acted upon, and may result in outputs (first-order, intentional consequences) or outcomes (second-order, unintentional or unforeseen effects). Policies have five components which result in a *pattern of behaviour*:

- an object or set of objectives
- a desired course of events
- a selected line of action (or inaction)
- a declaration of intent
- an implementation of intent.

Thus, a policy can be 'viewed as a pattern of purposive or goal-oriented choice and action rather than as separate discrete decisions' (Mitchell, 1989: 263–4). That is, a policy is a response to an issue resulting in a consciously chosen course of action (or inaction) directed towards some end.

Fundamentally, the key issues are those first posed by Lasswell (1950) for the study of politics, namely: who gets what, when and how? The answers to these questions require a consideration of power, conflict and ideology within the decision process of resource management. This chapter commences by examining these various perspectives on public policy. A systems model of the policy process is then outlined. This model suggests that public policy is the result of a process involving interest representation, decision making and administration. Using the model as an organizational guide, the chapter explores the implications for impact assessment arising from the public policy process. First, the role of public administration in the formation and implementation of policies is considered. Various approaches to understanding environmental decision making are then outlined. Lastly, the chapter considers how stakeholders' interests are represented within the policy process by outlining the dynamics of interest representation.

Public policy perspectives

Public policy may be defined as: 'A set of interrelated decisions taken by a political actor or group of actors concerning the selection of goals and the means of achieving them within a specified situation where these decisions should, in principle, be within the power of these actors to achieve.' (Jenkins, 1978: 15). This general definition of public policy has several important dimensions (Wilson, 1981). It indicates that decision making is crucial to policy making but that it is only one aspect of the policy process. The idea of inaction as a decision (and the allied concept of a non-decision) is left open and the need to understand the institutional setting for policy making is highlighted. Lastly, the definition makes the conceptual distinction between *policy content* (the substance of the policy itself) and the *policy process* (the strategies, techniques and methods of policy making).

Noting that policy study had become focused on the study of bureaucracy and public administration, Simeon (1976: 549) has argued that policy making 'must be broader than public administration' and that 'a broader political framework' is necessary for policy studies. A political framework provides a series of constraints and opportunities that define the determination of problems, acceptable policy responses and the procedures for their evaluation. A range of factors is involved, including the social and economic environment, the system of power and influence, the dominant ideas and values in society, and formal institutional structures. The policy process reflects and is shaped by this broader political framework (Simeon, 1976).

Three dimensions of public policy are fundamental: the scope of government policy, the means used to assure approval or compliance with decisions and the distributive dimension or 'who gets what?' (Simeon, 1976: 559). It is essential to view policy as an 'interplay between ideas, structures and processes' stemming from the main features of public policy and the policy system (Doern and Phidd, 1983: 34). These include:

- normative intent and the expression of ideas, values and purposes
- the exercise and structuring of power, influence and legitimate coercion
- dynamic processes of policy development
- the implementation of desired behaviour
- a series of decisions and non-decisions.

The normative content of public policy is a function of ideology, institutions and dominant ideas. For example, in Canada there has been an enduring concern for efficiency, individual freedom, equity, stability, redistribution,

51

national unity and identity, and regional sensitivity (Doern and Phidd, 1983).

Public policy involves some degree of legitimate coercion. It is a product of the need to structure and organize the power and influence of politicians, administrators and other stakeholders. The right to exercise those powers in society is usually accorded to the state (the sum of the legislature, executive, bureaucracy, courts, police and military authorities) and there are 'widely varying views about what the role of the state is and should be' (Doern and Phidd, 1983: 35–6).

The role of the state and the function of public policy can vary greatly from nation to nation. Numerous attempts have been made to classify the various types of government and political divisions in the world. The usefulness of frequently used classifications and typologies of contemporary political systems has been reviewed by Bebler and Seroka (1990). Most classifications have an inherent bias towards an ideological division into Western democratic states, communist totalitarian states and the Third World. However, progress beyond this rudimentary and unsatisfactory level has been stalled by the extent to which there is disagreement and uncertainty as to how to classify political systems and as to the nature of the key characteristics within any typology. As the work of Lijphart (1990) has shown, even democratic political systems offer a range of diversification that challenges successful categorization.

This caveat aside, it does seem important to attempt some clarification as to the basic political philosophy within which the present text is grounded. Hague and Harrop (1987: 45–68) examined world political systems distinguishing between: liberal democracies, communist party states, competitive oligarchies, military regimes and populist-mobilizing regimes. The dominant characteristics of liberal democracies were identified as:

- *Limited government*: majority rule is balanced by constraints on the exercise of power.
- *Constitutional government*: majority rule is qualified by due regard to the acknowledged rights of individuals and minorities.
- *Pluralistic government*: majority rule operates alongside numerous interest groups which are consulted about and bargain over government proposals which affect them.

The present discussion is consistent with this political philosophy and its discussion of public policy reflects the traditions of pluralistic Western societies rather than any Marxist, military, oligarchic or one-party political traditions or philosophies.

In order to understand variations in public policy, five general approaches can be identified (Simeon, 1976: 566–80). Public policy can be viewed as a consequence of:

- *the environment*: patterns of policy may be interpreted by reference to such broad environmental factors as demography, geography, levels of urbanization, wealth, industrialization, etc. Environment is probably more important as a contextual factor than as a dependent variable in any policy analysis.

- *The distribution of power*: patterns of policy will reflect the distribution of power and influence in society. Competing modes of explanation include those of élitism, pluralism and class analysis. All three approaches have merit but all three also have significant limitations.
- *Prevailing ideas*: emphasizes the cultural and ideological factors of dominant ideas, values and societal beliefs. Obviously central to the substance of policies and the means by which they are derived, ideas do not by themselves provide a complete explanation of policies.
- *Institutional frameworks*: the policy consequences of the institutional structure of the political system, its formal rules and regulations. This approach appears to be inextricably linked to the other approaches, and the institutional arrangements themselves may be viewed as policies.
- *The process of decision making*: closely related to institutions and the focus of the majority of policy studies. This approach examines the ways in which stakeholders interact with one another in the making of policy.

These perspectives can be integrated within a systems model of the policy process (Fig. 4.1). Most policy process models have their basis in the work of David Easton and his concept of a dynamic model for the political system, wherein politics was defined as the authoritative allocation of values. Easton's use of the system's paradigm has been credited with moving political science away from an exclusive concern with government institutions and towards the relationship between government and society (Hague and Harrop, 1987: 22–4).

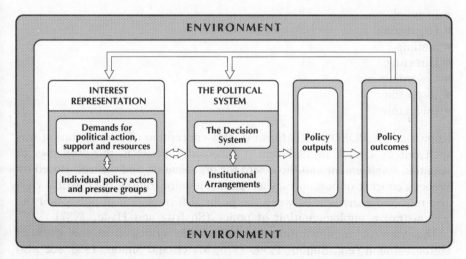

Fig. 4.1 A systems model of the policy process

The basic Easton model stressed the 'relationship between socioeconomic inputs (the forces), political system characteristics (systems) and policy outcomes (responses)' (Wilson, 1981: 164–5). In an amended version of the systems model presented by Jenkins (1978: 22), the inputs were further broken into demands that were expressed in the political system by mediating variables (groups, parties and organizations, etc.) and the political system itself was divided into a decision system and the 'organizational network'.

Figure 4.1 is a further iteration of the systems model of the policy process. It reflects both Jenkins's modifications and the description of Easton (1965). The model suggests that policy results from a process involving:

- *Interest representation*: policy inputs take the form of demands for political action, support and resources that are expressed by individual policy actors and pressure groups.
- *The political system*: authoritative rather than routine decisions are reached that reflect the decision-making process and the institutional arrangements that structure the administration of policy.
- *Outputs*: this system acts to produce policy results that are intended (outputs) and impacts that were not (outcomes). These provide feedback within the system, prompting the initiation of further policy responses.

Administering policy

The central concern of public administration is the authoritative allocation of values through public organizations (Wilson, 1981: 11). The classical formulation held that administration involves seven major functions:

- Planning
- Organizing
- Staffing
- Directing
- Co-ordinating
- Reporting
- Budgeting

Known as the POSDCORB formula, this conception of administration was spelt out by Gulick in 1937 and it reflected a belief in the principles of scientific management and the desire for organization and control. Based on Weber's concept of bureaucracy as rational administration, the classical view also reflected a division between the political formulation of policy and the administrative implementation of policy (Shafritz and Hyde, 1987).

This view of administration was thoroughly criticized in the work of Herbert Simon (e.g. Simon, 1959, 1976; March and Simon, 1958; see also Wilson, 1981). Simon believed that administrative organizations were primarily decision making structures and that decisions were of two fundamental types. Decisions were either based on *factual premises* (descriptive of what is) or on *value premises* (prescriptive of what should be). He rejected the classical notion of 'economic man' (the administrator who is rational, has full information, searches for optimum solutions and maximizes every decision) in favour of 'administrative man' (the decision maker who operates within the confines of 'bounded rationality', has incomplete information, looks for the most acceptable solution and 'satisfices' by accepting decisions that are good enough). Simon established the basis for decision theory and the contemporary focus on the policy process. Within this approach to public policy, the view may be taken that 'policy *is* administration and that administration and management are inextricably linked' (Wilson, 1981: 11).

Policy is not static. Moreover, at any given point 'it may not be possible to say whether action is influencing policy or policy action' (Barrett and Hill, 1984: 219). Thus, the classical differentiation between the political formulation and administrative implementation of policy appears anachronistic and a more accurate portrayal holds that 'implementation must be regarded as an integral part of the policy process rather than an administrative "follow on" from policy making' (Barrett and Hill, 1984: 220).

This view of the policy process is consistent with that of Gawthorp (1983: 122) who described the public policy process as 'the levels of recursion that flow from normative policy pronouncements to specific organizational strategies (programs), to rational organizational designs (structures), and, finally, to explicitly articulated organizational operations (procedures).' Policy making involves change and change is 'an inherent and ever-present element of every organization' engaged in policy making (Gawthorp, 1983: 120). There are two organizational choices in responding to change: change organizations or organize for change (Gawthorp, 1983). The decision to organize for change entails:

- *Organizing for control*: this approach reflects Weber's basic tenet for bureaucracy and suggests that control is best achieved through the use of a vertically structured, hierarchical organization.
- *Organizing for bargaining*: where political efficacy is the primary goal, the organization must allow for a 'wide range of reciprocal exchange interactions', i.e. administrators function as brokers of information and data who develop a consensus with the other actors in the policy system.
- *Organizing for responsiveness*: the organization accentuates its accountability by viewing change as value enhancing. Change is actively sought rather than avoided and the organization is structured around independent project team clusters and an open decision architecture.
- *Organizing for analysis*: the basic designs of systems theory are utilized to

structure decision making on the basis of rational, comprehensive and comparative analysis.

Gawthorp's work suggests that there is no one single approach to organizing or reorganizing to accommodate change. Each of the approaches identified is complementary and different approaches may be used by different organizations at differing times or at the same time. The net result is a further layer of fragmentation in the operating environment of organizations (Gawthorp, 1983).

Policy analyses must account for differing institutional settings. It is important to recognize that the 'policy process is dynamic but disorderly' and that as issues change they follow different sequences (van Horn, Baumer and Gormley, 1989: 25). The implication is that policy cannot be reduced to orderly steps. Moreover, since issue characteristics constrain policy makers in significant ways, it has been stated that there is 'no single policy process but several and that politics is a significant determinant of policy outcomes' (van Horn, Baumer and Gormley, 1989: 26). Based on this rationale, van Horn, Baumer and Gormley (1989) linked politics and policy together, presenting a range of alternative 'images' of the policy process:

- *boardroom politics*: private sector decision making by business élites and professionals that has important public consequences
- *bureaucratic politics*: administrative rule making and adjudication by government bureaucrats, with input from client groups and professionals
- *cloakroom politics*: legislative politics, with policy making by legislators constrained by various constituencies
- *chief executive politics*: policy making by key legislators, a process dominated by presidents, governors, mayors, and their advisers
- *courtroom politics*: judicial policy making through court orders, in response to interest groups and aggrieved individuals
- *living-room politics*: the political galvanization of public opinion, usually through the mass media.

These six images were used by van Horn, Baumer and Gormley as a framework to analyse the policy process. The images also functioned as bridge between issue characteristics (such as salience, conflict, complexity and costs) and policy consequences (change, responsiveness and other outcomes), serving to demonstrate that non-traditional areas of policy making deserve attention (van Horn, Baumer and Gormley, 1989).

This use of imagery reflects the fact that there are vastly differing notions about the decision-making process and the roles of stakeholders involved in the determination of public policy. As the images suggest, these differences extend from variations in political philosophy, through concerns for legitimacy, accountability and efficiency, to uncertainties as to the ways and means by which interests should be represented in decision-making processes.

Environmental decision making

There are many different models of the decision-making process and the literature in this field has been reviewed at length by Mitchell (1989: Ch. 12), Doern and Phidd (1983: Ch. 6), Wilson (1981: Ch. 6) and Jenkins (1978). Decision making models may be classified in several different ways, but the most basic distinction remains that between *prescriptive* and *descriptive* decision-making models. Prescriptive models attempt to outline how decision making should occur. Conversely, descriptive models seek to record how the decision process actually functions in practice. The rational comprehensive model characterizes the former, while the latter are best represented by the concept of disjointed incrementalism.

In this instance, the prescriptive/descriptive typology is especially valid as there have been few attempts to analyse resource management or environmental planning situations using models that do not fit or conform to this categorization. For example, neither the public choice approach nor the class analysis model identified by Doern and Phidd (1983) have seen much application in the context of environmental decision making. Furthermore, while such models as Lowi's input–output approach, Etzioni's concept of mixed scanning and the concept of the appreciative system developed by Vickers have their adherents (Wilson, 1981), they have been infrequently employed in the analysis of resource policy. Thus, the rational comprehensive model and disjointed incrementalism can be employed to provide a useful contrast in the characterization of environmental decision making.

The rational comprehensive model

The rational comprehensive model is the best known prescriptive model of decision making. Based upon a presumption of *rational* behaviour, the concept of *economic man* and the use of the scientific method to derive optimum (correct) solutions, the rational comprehensive model seeks to prescribe how decision making should occur to provide for systematic planning (Mitchell, 1989: 264; Doern and Phidd, 1983: 139). As shown in Fig. 4.2, the rational comprehensive model involves several sequential steps:

- *Identify the problem*: the problem is defined and isolated from other concerns.
- *Determine goals and objectives*: the values, goals and objectives of the decision maker are identified and may be ranked so that the desired results (or 'ends') are known.
- *Review alternative strategies*: the ways and means of achieving the ends are determined by identifying the range of possible solutions and their respective benefits and costs.
- *Estimate potential impacts*: the consequences of each alternative strategy are assessed.
- *Select the preferred plan*: the various alternatives are compared and the alternative that maximizes net expectations is selected.

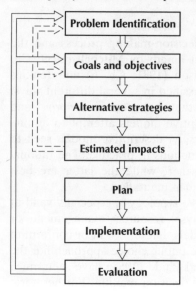

Fig. 4.2 The rational comprehensive model of decision making

- Implementation: the preferred plan is put into operation.
- *Evaluation*: the success of the plan is determined and behaviour modified as appropriate to correct errors.

The rational comprehensive model represents 'an ideal pattern of choice and activity' (Mitchell, 1989: 265) which is rarely achieved in practice. Absolute rationality is a myth. Decision makers usually operate in conditions of imperfect knowledge and uncertainty. A full range of alternatives is assessed only on rare occasions and 'correct' solutions are an illusionary goal in most environmental situations.

What then is the value of the rational comprehensive model? Its utility is that it provides a normative standard against which many policies and decisions can be assessed and it 'provides a code of language around which political and policy debate is engaged' (Doern and Phidd, 1983: 140). The rational comprehensive model does not have to be realized in practice: its value is that it serves as a commonly recognized departure point in decision making.

Disjointed incrementalism

The concept of disjointed incrementalism is derived from the work of Lindblom (1959, 1979; Braybrooke and Lindblom, 1963). It is a descriptive model of decision making that builds upon criticisms of the rational comprehensive model and embraces Simon's notions of *bounded rationality* and *satisficing*, to outline how decision making occurs in practice. Incrementalism suggests that decisions are made in small sequential steps in which there is no grand strategy. Therefore, organizations do not undertake systematic planning but act to resolve problems, a feature Lindblom termed the 'science of muddling through'.

Disjointed incrementalism has the following characteristics:

- Changes are small and incremental.
- Only a restricted number of choices are considered.
- Choices reinforce the status quo.
- Means and ends are reciprocal: problems are continuously redefined and changes are made to objectives so that they conform to the means by which they may be attained. Desired results are set on the basis of what is achievable rather than on what might be desirable.
- Action is directed towards remedying a negative situation rather than reaching a preconceived goal.

This description of the decision making process suggests a decision environment in which problems are not clearly defined, goals and objectives are often in conflict, the range of alternatives is very narrow and only 'significant' consequences are identified for alternatives (Mitchell, 1989: 266). Incrementalism suggests that no correct nor right solutions to problems exist. Rather, the emphasis in decision making is placed upon avoiding mistakes and/or avoiding the 'wrong' decision.

Disjointed incrementalism provides an accurate description of decision making in many instances but it fails to explain sudden or radical shifts in policy. The model places a high value on the need for consensus and agreement. It is more important that decisions be acceptable than correct and the mechanism of small, sequential increments from the status quo exists as a strategy to avoid mistakes. Under incrementalism, analysis and evaluation occur in sequence and decisions comprise a long series of amended choices. But the model fails to specify what constitutes an incremental intervention and provides no guide to decision makers on when they should intervene in a situation (Wilson, 1981). Thus, while disjointed incrementalism is 'intuitively persuasive at a commonsense level', it fails to explain decision making nor indicate how decisions should be derived (Doern and Phidd, 1983: 142).

Key concepts

The rational comprehensive model and disjointed incrementalism are best thought of as the opposite ends of a spectrum of models designed to explain the decision process. The normative ideal for decision making is represented by the rational comprehensive model, while the reality that is often decision making practice is described by disjointed incrementalism. Both models have their value and, as shown in the review by Mitchell (1989: 269–82), they have infused studies of environmental decision making.

In applying general decision making models to environmental decisions, a number of mutually supportive concepts have emerged. The first of these is the separation of routine and strategic decision making (Sewell, 1974). *Routine decisions* do not involve resource conflicts, whereas *strategic decisions* encompass fundamental allocative conflicts and invoke a more politically conscious course of policy determination. The motivating force delineating strategic decision making is the allied notion of *stress* (Kasperson, 1969; Wood, 1976). Stresses on the policy environment stem from competing demands on scarce resources. Consistent with the systems model of the policy process, stress is articulated to decision makers by pressure groups or the media. Decision makers respond to stress through incremental decisions designed to provide an adjustment to new political realities. Lastly, O'Riordan (1976, 1981) has advanced the idea of *decisional pathways* to describe how decisions are made in response to stress in strategic decision making. Based on the premise that 'most decisions take place in response to political stress initiated by pressure groups' (O'Riordan, 1981: 245), the results can either be:

- *non-decisions*: either through the symbolic use of power or the suppression of political debate
- *routine decisions*: conventional problem-solving actions, or
- *strategic decisions*: either through a crisis response, such as the political placation of pressure groups, or participatory negotiations and an explicit attempt to incorporate the views of the public (O'Riordan, 1976).

O'Riordan's approach implies that decisions arise from a process of incremental bargaining among stakeholders. There is extensive support for this characterization of environmental decision making, which explicitly recognizes that resource management is, in essence, *a bargaining process*.

In one of the best expositions on the subject, Dorcey (1986) outlined the governance system for resources on the basis of four key characteristics:

- A wide range and number of public and private actors are involved.
- Individuals and groups act through loosely coupled organizations.
- These interactions create numerous arenas for decision making.
- The actors, organizations and decision-making arenas 'are linked by the common thread of bargaining' (Dorcey, 1986: 63).

Adapting the definition of Fisher and Ury (1981), Dorcey (1986: 4) defined bargaining as: 'back and forth communication designed to reach an agreement when two or more participants have some interests that are shared and others that conflict'.

Bargaining is necessitated because conflicts have to be resolved and resolution involves compromise in the accommodation of different interests. Bargaining is between and within public and private interests. Moreover, since competition for resources is increasing, so is the level of resource conflicts. Concomitantly, opportunities for unilateral solution through co-operation, accommodation and dictation are diminishing. Hence there is a need for jointly derived solutions and greater attention to bargaining as the process for achieving compromise between policy actors (Cormick, 1980).

The success of bargaining is influenced by institutional structures, the power and influence of the actors involved, their individual bargaining skills and the opportunities for co-operation and accommodation (Dorcey, 1986). For bargaining processes to work well it is important that participants be well informed and that the affected interests (the stakeholders) should have the opportunity to participate. In addition, successful bargaining focuses on the principles of each stakeholder, rather than their avowed positions at the onset of negotiations (Fisher and Ury, 1981).

Failures in bargaining result when either: (1) bargaining fails to occur, or (2) the bargaining process itself fails. Bargaining fails to occur when there is a refusal to bargain by one of the parties involved, when there is an absence of a suitable bargaining forum or when stakeholders are excluded or omitted from a bargaining forum. Failures in the process of bargaining itself result when the nature of the conflict is misunderstood and advocacy acts to increase rather than diminish conflicts, or when talks break down owing to inept bargaining (although mediators and a process of arbitration can be employed to help circumvent these eventualities).

Successful bargaining also necessitates that the basis of *conflict* be understood and its different manifestations be recognized. Environmental conflicts arise out of differences in technical judgements (*cognitive conflicts*), from differences in ideology (*value conflicts*) or as a result of differences in overt behaviour (*behavioural conflicts*). They may be manifest or latent in form and may arise at differing stages during problem solving (Dorcey, 1986; Vari, 1989; Bergstrom, 1970). The most crucial conflicts are those that arise from a state of tension in situations of incompatibility. Indeed, a conflict of interest may be defined to exist between two parties 'if, and only if, their interests are incompatible' (Bergstrom, 1970: 204).

Environmental conflicts have several important distinguishing characteristics (Bidol and Lesnick, 1984). They are scientifically complex and often involve irreversible effects, considerable uncertainty and ideological differences. Normally, environmental conflicts are typified by an inability to determine precise boundaries, parties or costs. Consequently, they can include many stakeholders who may emerge at different times during the dispute and they severely strain existing decision-making processes.

In theory, the resolution of conflict can involve either the elimination of parties, reconciliation or the elimination of incompatibility (Bergstrom, 1970). Assuming that the option to eliminate parties can be rejected, the resolution of environmental conflicts emphasizes reconciliation and the elimination of incompatibilities. These objectives may be achieved through confrontation, appeasement or bargaining (Vari, 1989). In each instance, the process of conflict resolution must account for the relative power of the parties involved and the degree of trust existing between them. For example, bargaining functions best in mixed power systems where parties exist as equals. It has less utility in either a unilateral power system (one party is dominant), or in a bilateral power system (party A plus party B are equal to party C), where the incentives to bargain are largely absent (Bonoma, 1976).

Dorcey (1986) maintained that there is a need to consider the design of bargaining processes and their integration with authoritative decision making structures, especially with respect to such issues as the legitimacy, accountability and representativeness of bargaining. The premise of environmental decision making as a bargaining process conforms to the notion of policy implementation as a 'political process characterised by negotiation, bargaining and compromise between those groups seeking to influence (or change) the actions of others, and those upon whom influence is being brought to bear' (Barrett and Hill, 1984: 238). This view of policy draws together the ideas of bargaining, incrementalism and pluralism, and incorporates them into an integrative framework that presents 'a working formula in which the incremental allocation of limited resources within a pluralist society is accomplished by a closed group of professional political participants on the basis of bargained political agreements' (Gawthorp, 1983: 126).

Decision making and public policy should not be viewed narrowly. Rather, the policy context and the role of conflict, consensus, power and the interests of stakeholders must be inherent in any analysis. Concomitantly, this perspective also focuses attention on how the various interests in resource issues should be represented in the policy process.

Interest representation

Conflict is inherent within environmental planning. The nature of the resource base, competing and often incompatible uses, and the wide range of vested interests ensure the potential for disputes in all allocative management decisions affecting the environment. To be effective, environmental decision making must provide for bargaining among the interested parties involved. Smith (1988b: 161) has suggested that decision processes should be based upon:

- real and regular consultation
- a common data base
- action plans involving multiple stakeholders
- a variety of flexible mechanisms.

The difficulty is in finding the means to successfully identify, address and incorporate the affected interests into decision making. Three main approaches to interest representation within decision making and the policy process for resource management have been utilized: lobbying, public participation and environmental dispute resolution.

Pressure groups and lobbying

Lowe and Goyder (1983: 2) defined pressure group activity as the 'efforts of organized groups to influence the decisions of public authorities'. Such groups may be communal, customary, institutional, protective, promotional or associational in character (Hague and Harrop, 1987: 121–5). Irrespective of their basis, pressure groups are 'organizations whose members act together to influence public policy in order to promote their common interest' (Pross, 1986: 3). Pressure groups are characterized by:

- persuasion
- organization and continuity
- articulation and aggregation of a common interest
- the desire to influence power rather than exercise responsibility for government.

They attempt to influence public policy by either swaying public opinion (through media usage, advocacy advertising or political contributions) or through the practice of lobbying. Pressure groups are an integral feature of the political system and it is important to recognize that 'all societies do exhibit some degree of interest articulation' (Hague and Harrop, 1987: 114). Pressure groups form part of the linkage between public officials and private citizens. However, their existence and the activity of lobbying generate much debate concerning the issues of accountability, equity, secrecy and the function of interest representation.

For example, within pluralist political theory, freely organized interest groups can be seen as 'intermediaries in a two-way flow of communication between rulers and the ruled' (Hague and Harrop, 1987: 115). Political pluralism and the 'group theory of American politics' were developed though the 1950s and 1960s into a dominant normative theory for politics in the United States. The theory was based upon empirical studies which indicated that few individuals have access to the policy process. Instead, most people rely upon organizations and pressure groups to express the need for new policy initiatives (Cayer and Weschler, 1988). Pluralist democratic theory postulates that pressure groups play a central role in the policy process, seeking to influence governance through lobbying. Meanwhile, government officials act as brokers attempting to achieve compromises and trade-offs between the competing interests.

This conceptualization of democracy has its flaws and has been subject to substantial criticism (Hague and Harrop, 1987). In its basic form, pluralist theory fails to take into account that government agencies themselves often act as interest groups. The basic form also implies that government is merely a captive element of pressure group politics, a view that has been rejected 'as too extreme because it ignores the independent role which the state plays in deciding which interests are satisfied and which are not' (Hague and Harrop, 1987: 114).

Rather, it is more productive to focus on the relationship of interest groups to political power and to view pressure groups as a constraint within the structure of policy making (Sandbach, 1980). This perspective has been used by Pross (1986) in his development of the concept of *policy communities*: that part of the political system which by virtue of its functional responsibilities, vested interests and specialized knowledge, acquires a dominant voice in policy determination. There are two segments within a policy community: the sub-government and the attentive public. In effect, the sub-government (including executive agencies, legislative committees and government agencies) acts as the policy making body that processes most routine decisions. The attentive public (agencies, organized interests, political parties, media) is neither as tightly knit nor as clearly defined. These groups seek to influence public policy but do not participate in the policy process on a regular basis (Pross, 1986). The goal of pressure groups is to move beyond their role as an attentive public advocating change, into a position as participants in the formulation of policy by the sub-government (Coleman, 1985).

The central strategy utilized by pressure groups to gain influence in the policy process is lobbying. Lobbying can be defined as 'the stimulation and transmission of a communication by someone other than a citizen acting on his own behalf, directed at a government decision maker with the hope of influencing his decision' (Milbraith, 1963: 8). Stanbury (1986) has analyzed the practice of lobbying on the basis of:

- *Timing*: when in the policy process should lobbying occur?
- *Targets*: who it is a group is attempting to influence.
- *Vehicles*: the mode of communication employed.

His work suggests that inputs must be made to the policy process as early as possible. How those inputs should be made and to whom, will vary from country to country. In Canada, Stanbury has indicated that the preference for successful lobbying is for an informal and unrecorded meeting with a senior policy adviser within the government bureaucracy. These discussions are subsequently pursued, reinforced and legitimized by the submission of a formal brief to the relevant government agency (Stanbury, 1986). In contrast, most 'public' interest groups attempt mass campaigns directed at elected politicians. Not only is this an inefficient vehicle for lobbying but it is also directed at a target whose role in policy deliberations is relatively minimal in comparison to that of other policy actors (such as the executive or cabinet, their support staff and senior agency officials).

The key factor in successful lobbying is access and 'the ability to gain entry to key decision-makers to make a representation' (Stanbury, 1978: 185). In addition, successful lobbying requires financial resources, an ability to deliver political benefits (such as votes) and clearly defined policy alternatives. Moreover, since well-established pressure groups and corporations have these attributes and most environmental interest groups do not, the practice of lobbying is rarely equitable (Smith, 1982a).

Thus, these factors can also be used to distinguish between different interest groups. Hague and Harrop (1987) have suggested that the key dimensions to categorize groups are:

- *access*: direct contact with government, indirect influence (on political parties or through the media), protest and direct action.
- *the determinants of influence*: the nature of the system itself, legitimacy, available sanctions and tangible resources.

Similar characteristics were used by Lowe and Goyder (1983) who presented a comprehensive analysis of the character and role of environmental groups in Britain. Their work traced the development of the contemporary environmental movement and the emergence of environmental interest groups as a political force. Throughout, Lowe and Goyder (1983) focus on the internal organization of interest groups and their external relations with the political system.

More generally, Pross (1986) has suggested a typology which classifies pressure groups from issue-oriented to institutionalized. The typology is based upon group characteristics (objectives and organizational features) and their levels of communication with government (media-oriented and access-oriented). Coleman (1985) extended this concept one step further by distinguishing between the role of groups as *policy advocates* and as *policy participants*. Pressure groups seek to move from a position of advocacy to one of policy participation. Coleman identified this transition to be a function of a group's ability to balance the pressures for co-ordination with the need to retain autonomy. That is, a pressure group must balance its efforts to achieve legitimacy within the power structure of the policy community, with the interests of its membership and their desires for independent advocacy. The former provides the group with increased influence and access to policy participation, but the latter is crucial to the group's image of independence and the maintenance of its resource support (Coleman, 1985).

Most empirical studies have emphasized the mobilization of resources as a good predictor of interest group influence and success. This aspect of group politics has been severely criticized from a class perspective by Sandbach (1980). He maintained that as resources are not evenly distributed within societies, pluralist theories that assume equal access to resources should be rejected. His solution was to advocate that decision making be as open as possible and for it to 'at least have the appearance of being representative' (Sandbach, 1980: 114).

Pressure group politics threaten the legitimacy of the state when they undermine the understanding of the legislature as the means of governmental accountability (Pross, 1986). The policy community is not representative of the public and when it is viewed as a 'functional constituency' it diminishes the legitimacy of the legislature. The solution is to build more equality into the political system through the creation of 'safeguards against the usurpation of legislative power by policy communities' (Pross, 1986: 211).

Lobbying can be an extremely effective means of influencing public policy. However, its effectiveness is dependent upon groups having the necessary skills, acumen and resources with which to lobby. Environmental groups do not normally have these necessary precursors and cannot, therefore, lobby on an equal footing with established pressure groups. Moreover, increased lobbying has an inherent tendency to 'narrow, rather than broaden, the base of the pluralist system' (Smith, 1982a: 564).

The key is the issue of representation. For the potential of group representation in the policy process to be realized, opportunities for a wider public to influence policy decisions must be expanded. What this means for interest representation is the development of mechanisms to balance the influence and effects of lobbying (Smith, 1982b).

Lay citizens and public participation

Public participation refers to a group of methods and procedures designed to consult, involve, inform and empower lay citizens and interested groups regarding environmental issues. Based on the moral recognition that those affected by a decision should have input to that decision, public participation may be defined as 'any action taken by an interested public (individual or group) to influence a decision, plan or policy beyond that of voting in an election' (Smith, 1984a: 253–4). Public participation can be seen to be a means to achieve democratic values, especially the representativeness and responsiveness of political and administrative decision making. Conversely, public participation is often viewed as a means to legitimize administrative decisions on the assumption that consultation will ease the implementation of policy and have a cathartic effect on dissenters (Smith, 1984a: 254).

Four prerequisites for effective public participation were identified by Smith (1984a) as being:

- the legal right and opportunity to participate
- access to information
- resource provision
- the representativeness of participants.

As outlined in Chapter 3, environmental law plays a pivotal role in providing legislated access to decision making processes. However, whereas legislation may contain provisions for public input, often they may be invoked only at the discretion of administrative officials. In practice, this has meant that most public participation experiences have occurred on site- and/ or issue-specific concerns, where a definable interest in property or rights can be more readily recognized (Smith, 1987a).

Interested publics are often the victims of misinformation, supposition and fear regarding resource development proposals. The only way to counter public uncertainty is for proponents to provide clear and accessible sources of information. This has not always been 'standard operating practice' for proponents or government agencies. Information has not always been available, nor has it been presented in a functional manner (i.e. free of bias and the perils of 'data overload' that result from a surfeit of data but the absence of any appropriate conceptual frame of reference). It is only through the advent of freedom of information statutes and judicial rulings that easier access to information has become more readily available to the interested public.

Public participation often proceeds over an extended time-frame, especially where large projects with many impacts are under consideration and/or in situations of impact assessment approvals by regulatory bodies. Continuity of participation, technical guidance, legal advice and attendance at participatory forums all necessitate financial, personnel and logistical resources. On occasion, funds for intervention have been provided to interested participants but the practice remains fraught with problems (e.g. who pays for intervenors? When should they receive funds? What criteria should be used in allocating resources? Should money be paid up front or only awarded as costs at the end of all proceedings?). The absence of adequate resources remains the most frequently cited reason for non-participation by the public in resource management.

Restrictions in opportunities for participation, information access and resources, collectively act to curtail widespread involvement by individuals in environmental issues. Rather, there is a reliance upon groups to represent the interested public in most participatory forums. This reliance raises concerns over whether or not the *public interest* is represented by the involvement of such groups, especially where attempts to secure a wider, lay involvement are not made or fail. One solution to this dilemma is to recognize that special interest groups 'do not represent the public interest, but it is in the public interest that they participate' (Berger, 1976: 5). Thus, the emphasis should be placed upon a balancing of the interests involved, rather than the number of the participants *per se* (Smith, 1984a: 255).

A wide range of techniques is available for public participation, including:

- information programmes
- social surveys
- open houses
- public meetings and forums
- public hearings
- advisory committees
- task forces
- seminars and workshops
- simulation exercises.

These techniques vary in a number of dimensions, including their *intent* (from opinion sampling, through consultation to joint planning), their

67

approach to information (from the transmission of information, through information collection to information sharing) and the *nature of the public they involve* (from a lay audience, through interested groups to specialized interests and individuals). Good participatory design necessitates that a blend of approaches be utilized that balances the intent of involvement with the role of information and the type of public to be involved. In practice, however, there has been a reliance upon public hearings as the primary means of public participation in resources management (Smith, 1984a, 1987a). Indeed, their use has been so prevalent that public participation is often incorrectly assumed to be synonymous with public hearings.

A public hearing is a formal forum where previously submitted briefs are orally presented, summarized and examined, usually by legal counsel. Consciously modelled on judicial procedures, hearings can serve a useful role in the adjudication and evaluation of opposing, well-established viewpoints. However, they are inherently adversarial and tend to polarize opinions rather than foster consensus. They are notoriously poor mechanisms for information exchange, do not facilitate dialogue and are poor evaluators of the 'public interest' (Smith, 1984a; Checkoway, 1981). Hearings continue to be used because they are perceived by agencies and proponents to be a quick, cheap and simply administered means to satisfy the legal requirements for public participation. Hearings enable administrators to control public participation, while providing an opportunity for public frustration to be aired, antagonism to be diffused and decisions to be legitimized.

Attempts to move public participation beyond an emphasis on the use of hearings on site- and issue-specific concerns initially centred on the perspective of evaluation. The focus on evaluation stemmed from a need to identify answers to the questions 'what had worked? where? and why?' relative to the experience with public participation in practice. If alternatives to hearings were to be advocated, there had to be some means of determining the relative effectiveness of differing approaches to public participation and the circumstances under which they should be employed.

Building upon a critique of other attempts at evaluation, Smith (1984a, 1987a) developed an evaluative schema for public participation. Under the schema, public participation is considered within three phases: the *context* within which participation occurs, the *process* by which involvement takes place, and the *outcome* of that involvement (Fig. 4.3). The context for public participation programmes is a function of the opportunities and barriers presented by the institutional arrangements for management, the historical development of the issues involved and the nature of the lead agency whose mandate it is to provide for participation. The participation process itself must take into account the different perspectives of the various participants and how that might influence their approach to the participatory programme. In addition, the prerequisites for effective participation must be addressed in terms of both the type of public involved and the methodology

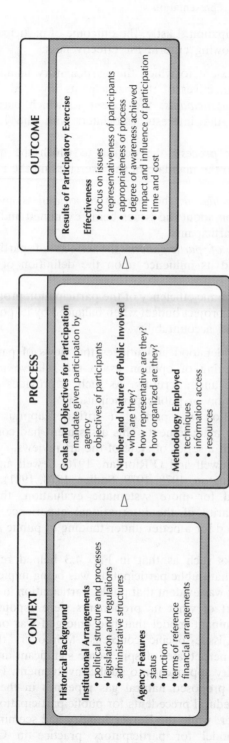

Fig. 4.3 A schema for the evaluation of public participation
Source after Smith (1984a, 1987a)

used to elicit their participation. Lastly, the outcome of participation can be evaluated using the following criteria for effectiveness:

- *focus on issues*: the extent to which the participatory mandate concurs with the goals and objectives of the participants involved
- *representativeness of participants*: the extent to which participation is representative of all the interests associated with the issues under discussion
- *appropriateness of the process*: the degree to which the methodology employed is suitable to the mandate of the exercise and the nature of the participants
- *degree of awareness achieved*: the amount of awareness and education created by participation about the issues being examined and the various perspectives of the participants
- *impact and influence of participation*: the effects of participation on eventual decisions and its influence upon the definition of subsequent issues and concerns
- *time and cost*: the economic efficiency of the participation programme as a proportion of the total project budget when balanced by its political utility in terms of equity and accountability.

The schema is easily understood and utilizes the views of participants to provide a balanced filtering of opinion as to the relative merit of the participatory programme under scrutiny. As such, it provides a systematic approach to the evaluation of public participation.

The focus on evaluation was part of a critical reappraisal of public participation in resource management. This assessment had commenced at the end of the 1970s with a series of state-of-the-art reviews, symposia and commentaries (Utton, Sewell and O'Riordan, 1976; Sewell and Coppock, 1977; Fagence, 1977; Langton, 1978, 1979; Sadler, 1978, 1981). In addition to identifying the need for more systematic evaluation, these reviews demonstrated the similarity of the international experience with public participation and the need for a better understanding of public participation in practice.

Evaluative frameworks such as that in Fig. 4.3 helped reveal several deficiencies in the way that public participation was being implemented. By the onset of the 1980s, it was evident that public participation in practice was falling a long way short of what its proponents were espousing (Smith, 1987a, 1990a). The dominant model that had emerged was one of public hearings on site- and issue-specific concerns. Moreover, this view of participation was being perpetuated despite some significant and successful experiments with contrary approaches to public involvement. For example, the Berger Inquiry on proposed natural gas pipelines in the Arctic had established several procedural precedents for public participation in Canada (Smith, 1987a). However, while the Berger Inquiry had seemingly acted to define a normative model for participatory practice in Canada, few programmes have been able to emulate its methods successfully nor have they been able to match its effectiveness.

Two factors help explain this trend (Smith, 1987a). First, experiments with innovative approaches to public participation created a heightened awareness among interested groups and the lay public of the true potential for public participation to influence environmental decision making. At the same time, these experiments indicated to administrators and politicians that public participation connotes a loss of control over the decision-making agenda. Effective public participation necessitates a broader perspective and a willingness to devolve the locus of power in decision making. As this fact became more self-evident, the interested public reacted with an enthusiasm that was infrequently embraced by administrators or project proponents.

Second, many sincere attempts to involve the public were thwarted by groups espousing the philosophy of NIMBY: the 'not in my backyard' syndrome. These groups were reacting to the perceived imposition of unwanted land-uses in their vicinity. Often, a NIMBY reaction was the product of a vocal (and occasionally, non-resident) minority whose opinion was not always representative of the wider community. Conversely, valid local concerns regarding impacts and the perceived risks associated with projects began being dismissed on the basis that they were only a NIMBY response and, thus, should be discounted by decision makers. In both instances, the NIMBY syndrome emerged as a dominant constraint on the practice of public participation: proponents eschewing involvement pro- grammes as a pointless expense, while NIMBY groups refused to participate less they add credence to a decision making process they perceived as biased against their interests.

Both of these trends helped contribute to a growing disillusionment with public participation through the 1980s. In response, attempts have been made to reconceptualize the role of public participation within environmen- tal planning (see Ch. 5). In addition, attempts have been made to counter some of the perceived deficiencies of public participation through the advent of alternative means of interest representation. Too often it appears, the practice of public participation has presented a 'forum for the expression of latent conflicts without providing the means for those conflicts to be resolved' (Smith, 1990a: 29). In response, the latest manifestation of attempts to broaden the basis for decision making in resource management has been the emergence of environmental dispute resolution.

Selected interests and environmental dispute resolution

Environmental dispute resolution (EDR) encompasses a group of approaches including negotiation, mediation, policy dialogues and other related consensus-building techniques. These approaches are predicated on the need to recognize the role and importance of bargaining in management and they seek to provide a means by which parties can resolve their differences

(Smith, 1990a). The various approaches differ but all involve a voluntary process of joint problem solving and bargaining (Fig. 4.4). Thus, EDR is distinct from litigation, arbitration and administrative procedures in that it stresses the attainment of consensus rather than the traditional adversarial process for conflict resolution (Carlisle and Smith, 1989; Amy, 1990).

The goal of EDR is to reach a mutually acceptable, jointly derived solution to problems (Bacow and Wheeler, 1984). All approaches to EDR involve an explicit process of bargaining that incorporates face-to-face communication. Proponents of EDR suggest that the process has several compelling benefits (Smith, 1988a), and it has been promoted on the basis that it:

- facilitates better problem identification and expansion

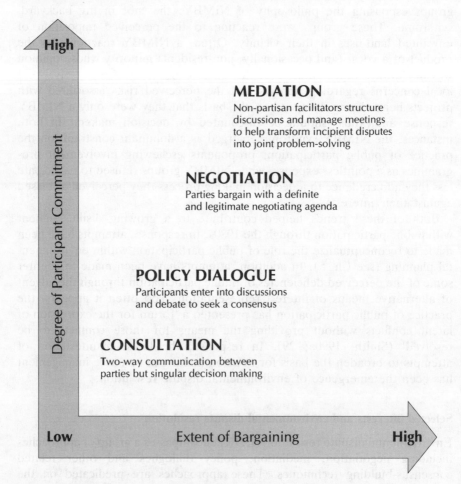

Fig. 4.4 Approaches to environmental dispute resolution
Source after Carlisle and Smith (1989)

- empowers a full range of stakeholders
- makes explicit and visible competing and conflicting interests
- aids in the implementation of solutions through the establishment of compromise and the accommodation of each stakeholder's interests.

It has also been promoted on the basis that it is not only faster and cheaper than litigation, but that it focuses upon the real issues in environmental disputes: 'to challenge projects in court, environmental groups must usually do so on procedural grounds. . . . While the real substantive issues in the dispute may be quite different – centering on the specific environmental impacts of the project – the only *litigable* issues are the procedural ones' (Amy, 1990: 217).

Critics of EDR contend that many of these supposed benefits are illusory. They suggest that the perceived weaknesses of litigation have been exaggerated, that EDR is biased in favour of corporate interests, and that EDR ignores the true basis of environmental conflict (Amy, 1990). These criticisms encompass the view that:

- The EDR approach is no faster nor cheaper than litigation in practice and it is only the threat of extended litigation that gives EDR the appearance of being more efficient.
- Only those parties with political power are invited to participate. Moreover, because substantial imbalances of power exist between stakeholders, EDR is biased in favour of those with greater access to expertise, information and financial resources
- The EDR approach is based on the premise that all environmental disputes are negotiable, which ignores any differences in fundamental values or environmental ethics that may exist among the different parties involved.

Thus, despite its increasing popularity throughout the 1980s, EDR is not without its own weaknesses. Moreover, because every dispute has different circumstances and conflict histories, normative models for EDR are optimistic at best and misleading at worst (Smith, 1988b). One of the underlying deficiencies of EDR has been the absence of any real models or directives for the successful resolution of conflict. There are, however, several guidelines and principles that all negotiating strategies share (see Fisher and Ury, 1981; Dorcey, 1986; Raiffa, 1982; Sullivan, 1984; Susskind et al, 1978). The decision-making process within EDR involves the following steps (McKinney, 1988: see also Bacow and Wheeler, 1984; Cormick, 1980):

- The various parties affected by, or interested in, an issue are included both in the process and in the determination of the nature of the process itself.
- The process stresses mutual education through non-adversarial dialogue and information exchange: assistance is provided to allow all parties to acquire the information and skills required to participate in bargaining.

- Options for mutually acceptable resolution of conflicts are developed and explored, and a third-party, external mediator may be employed to develop a negotiated solution acceptable to all parties.
- Decisions are made or influenced by the parties as they seek to derive a mutually acceptable solution that has provisions for implementation and subsequent monitoring and evaluation.

This emphasis upon consensus building and the involvement of all affected parties in the determination of decisions, are the major factors cited in favour of the EDR process (McKinney, 1988).

However, several problems remain unresolved by the adoption of EDR (Bingham, 1986; Jacobs and Rubino, 1987; Lake, 1987). In his summary of the literature, McKinney (1988: 337) presented six common concerns:

- problems of representation
- difficulties with setting an appropriate agenda
- obstacles to joint fact finding
- co-optation owing to differences in power and/or bargaining ability
- getting parties to observe their commitments
- barriers in monitoring and enforcing negotiated settlements.

In resolving these concerns, the increased use of EDR in the future depends upon the ability of decision processes to make four adjustments (Smith, 1990c; Carlisle and Smith, 1989):

- the creation of formal bargaining forums
- increases in the interactive skills of the participants
- changes in the philosophy of management
- increased information generation.

The creation of formal bargaining forums is essential to preclude situations where bargaining fails to occur because stakeholders lack the necessary infrastructure to initiate dialogue. The need for skills development reflects the tendency for new approaches to be introduced in resource management without the necessary training to enable their successful implementation. Similarly, skills development will be wasted unless it proceeds in conjunction with administrative and attitudinal changes in management philosophies that have a predisposition counter-intuitive to a negotiated approach. Lastly, information is essential not only to the process of negotiation itself, but also in the establishment of the legitimacy of the negotiated approach to the resolution of a dispute (Smith, 1990c).

Summary

Gormley (1989: 62) characterized interest representation as the 'ideal type' of catalytic control because it requires the bureaucracy to be responsive without predetermining the nature of that response. He suggested that interest representation should emphasize representation more than democracy, pragmatism rather than idealism, and a diffuse range of interests (Gormley, 1989). This view implies that interest representation should seek to work within the existing political system rather than designing alternatives to it, and is consistent with the model of environmental decision making described by disjointed incrementalism.

Disjointed incrementalism seems readily to embrace interest representation and other positive attributes of decision making. Thus, while it was originally conceived as a descriptive model, disjointed incrementalism has also been used as a prescriptive model for environmental decision making (Mitchell, 1989). However, neither incrementalism nor the rational comprehensive approach are a panacea and prescriptive models of environmental decision making need to indicate more fully the requirements for effective interest representation. To date, the experience with lobbying, public participation and EDR indicates that effective interest representation in impact assessment must address four issues (Smith, 1990a, c):

- *The determination of the public interest*: answers must be derived both for the nature of the public interest regarding a specific undertaking and the question of who represents that public interest.
- *The provision of equity and justice*: decision making must not solely expedite resource development efficiency but must provide both efficacy and accountability.
- *Design and training*: no one technique in isolation can accommodate all of the various publics that either should or wish to be involved. The proper mechanisms for representation must be utilized and the mechanics of their utilization be clearly understood by those using them.
- *Power*: the extent of political, legal and administrative commitment to sharing decision-making responsibility must be clearly understood and the ramifications acceptable to all parties.

As this chapter has shown, these issues are central to the realization of the political process and the formation of public policy in the determination of what governments do. How governments manage and plan is contingent upon this context. In the next chapter, the focus shifts to the specific role of impact assessment in environmental planning.

CHAPTER 5

Planning and the role of impact assessment

Introduction

Impact assessment is a component of the planning process for resource management. Unfortunately, by focusing upon the production of an impact statement, the dominant approach has tied impact assessment to a product-based view of planning. But planning is much more than just a physical plan. Indeed, in many circumstances, the process of planning is more important than the actual plan that results.

The need for a broader view of planning has been prompted by the growth of regulation and the increasing complexity of the approvals process for resource development. These factors require planning to be more comprehensive in scope and better co-ordinated (Noble et al, 1977). For example, Blacksell and Gilg (1981) called for a more integrated and co-ordinated approach to countryside management in Britain to avoid the separation between resource planning and development planning.

A more integrative approach to planning has also been prompted by the growth of uncertainty and the emergence of 'turbulent environments' (Lang, 1986b). Rising uncertainty necessitates that planning be more strategic in focus and interactive in its methods. However, most approaches to environmental evaluation adopt the stance that decisions can be made on the basis of 'a technocratic approach to planning that carries science well beyond the bounds of its ability to advance the wisdom of public decisions' (McAllister, 1980: 262). To avoid this trap, it is important that the role of planning be thought of in broad terms as a process of value-based choice rather than as a method *per se*.

A key aspect of planning is the idea of design and it is important to consider planning within a wider context of resource management as 'the initiation and operation of activities to direct and control the acquisition, transformation, distribution, and disposal of resources in a manner capable of sustaining human activities, with a minimum disruption of physical, ecological, and social processes' (Baldwin, 1985: 4). Or as:

76

a set of activities and procedures, properly seen as integral elements of the development process, aimed at ensuring that development activities affecting the environment:

- provide net benefits to society
- are sustainable
- allow for the continuation of valuable non-consumptive uses of ecosystems (Munro et al, 1986: 1).

This chapter reviews the meaning of planning. It commences by examining two commonly posed questions: what is planning? And, is there a theory for planning? The various approaches and normative models that have been developed to integrate impact assessment and the planning process are then outlined. In particular, the normative model for impact assessment developed by Whitney and Maclaren (1985) is detailed. The chapter concludes by contrasting the normative situation to a description of current practice in impact assessment, examining the constraints and requirements for impact assessment to provide for better decision making, environmental planning, impact mitigation and conflict resolution.

Defining planning

Typically, planning is defined in terms of its objectives. For example, Benveniste (1981: 140) used the definition that 'Planning is the elaboration of a set of related programs designed to achieve certain goals . . . [and the] . . . planning process is the set of interventions and other actions undertaken during the elaboration of a plan.' The planning process is then specified to include both technical aspects of data gathering, analysis and plan formation, and political dimensions of elaborating goals, assessing feasibility and forming support for implementation (Benveniste, 1981). However, not all people would agree with this conception of planning.

In a highly influential paper, Wildavsky (1973) discussed the elusive nature of planning by showing that planning can be all things to all people. For example, planning is often thought of in terms of future control and has been defined as 'the ability to control the future by current acts' (Wildavsky, 1973: 128). Consistent with this definition, planning requires the 'specification of future objectives' and 'a series of related actions over time designed to achieve them' (Wildavsky, 1973: 131). This is the approach used in Benveniste's definition above.

But planning can also be thought of in many other ways. It can be the application of causal knowledge. Planning is a form of political power. It can be adaptive, it is a process of goal-directed behaviour and it can be viewed as an intention whereby it has value 'not so much for what it does but for how it goes about not doing it' (Wildavsky, 1973, 139). Lastly, planning is an act of faith. It is often valued not for what it does but for what it symbolizes. This is the view of planning as rationality and the application of reason to social problems (Wildavsky, 1973).

Wildavsky's article was written with irreverence in order to highlight the very real difficulties that exist in developing a satisfactory definition for planning. Planning has a large range of connotations, and for this reason Bolan (1983) has argued that a single, general normative theory of planning is unlikely. Rather, he envisaged the practitioner as the theorist. That is, it is planners themselves who determine whether planning will be a technical, mechanistic function or a human-centred process of change (Bolan, 1983). However, the view that planning is what planners do is an over-simplification that many reject and an extensive literature has developed around the theme of planning theory.

Planning theory

The dominant theory of planning may be referred to as procedural. Its origins stem from the conceptualization by Geddes of planning as consisting of three steps: survey, analysis and plan (Fagence, 1977). Procedural planning theory is best exemplified by authors such as Faludi. In a highly influential text, Faludi (1973) envisaged planning as a means to promote human growth through the application of rational procedures of thought and action. For Faludi, this entailed the separation of planning as a process from the substantive content of the plan itself: hence, the referral to a procedural model for planning.

The procedural model of planning has also been called the 'classical model' of the planning process (Cayer and Weschler, 1988). It assumes a definition that planning 'is the process whereby proposals for the future are devised and assessed' (Cayer and Weschler, 1988: 94) and consists of a series of interrelated steps:

- the identification of needs
- the specification of goals and objectives
- the development of alternative means to attain each goal
- the estimation of the costs of each alternative
- the selection of the most promising alternative(s).

This is the conventional approach to planning and is also referred to as 'synoptic planning'. It is inherently rational comprehensive in perspective and, in many ways, it is indistinguishable from the rational comprehensive model for decision making that was reviewed in Chapter 4. It is also an approach that has been the subject of much criticism and controversy.

For example, Darke (1983) suggested that procedural planning theory is only a partial approach to the explanation of planning. The basis of his criticism was the point that a rational comprehensive approach to planning is inherently confined within a conservative political philosophy that views social action from a strictly functional perspective. It does not, therefore, consider the role of the state as an agent of change, nor does it accommodate analysis on the basis of class, gender or political ideology. Moreover, procedural planning theories have a limited methodological basis, and are constrained by their adherence to the notion of rationality. Darke (1983) maintained that these failures prevent procedural planning theories from addressing adequately the larger social milieu of a planning endeavour and he argued against the separation of planning substance from the process of planning.

Darke's criticisms are not an isolated example and several alternative theories of planning have been developed to counter the perceived failings of the procedural model (Benveniste, 1989). In one of the better reviews of planning theories, Hudson (1979: 387) defined planning as 'foresight in formulating and implementing programs and policies'. He then presented a classification of planning traditions by identifying five schools of planning thought:

- *Synoptic*: the basic model of planning, synoptic planning has its basis in rational comprehensiveness. In essence, the other traditions all take as their point of departure the perceived limits of the synoptic approach.
- *Incremental*: the incremental model of planning builds upon the work of Lindblom and the concepts of bounded rationality and satisficing. It stresses the role of bargaining processes in the sequential adaptation of small changes favouring the status quo.
- *Transactive*: the transactive approach envisages planning as a process of mutual learning, with ideas and action evolving in conjunction with one another. Educational principles and opportunities for information exchanges between stakeholders are central elements of transactive planning models developed by such people as Friedmann (1973).
- *Advocacy*: advocacy planning is grounded in the politics of protest and confrontation practised at the end of the 1960s. The approach is characterized by the development of plural plans as opponents of proposals develop their own alternative plans rather than seeking to amend those of the proponent.
- *Radical*: radical planning builds upon humanist philosophy in the development of collective action, activism and personal growth (e.g. Friere, 1970). It also encompasses Marxist and other forms of critical analyses emphasizing a holistic approach to the understanding of the effects of class structures and economic relations.

Hudson (1979: 388) recognized that 'planning covers too much territory to be mapped with clear boundaries' and, instead, focused his commentary upon the central propositions of each planning tradition. He then compared the five traditions according to the criteria of the public interest, the human dimension, feasibility, action potential, substantive theory and self-reflectivity. He concluded that no one planning style can be effective on its own and that all of the traditions require parallel inputs from other traditions: 'The synoptic planning tradition is more robust than others in the scope of problems it addresses and the diversity of operating conditions it can tolerate. But the approach has serious blind spots, which can only be covered by recourse to other planning traditions' (Hudson, 1979: 396).

Hudson's succinct analysis acts as an effective limit to the debate over the veracity of particular planning theories. Suffice it to say that the search for various explanatory theories is indicative of the fact that planning is in transition (Checkoway, 1986). Current planning practice must adjust to changing political climates and the rising level of uncertainty referred to at the opening of this chapter. As a consequence, planning theorists have made several attempts to develop ways to link strategic thought and action to implementation in planning (e.g. Faludi, 1987; Wyatt, 1989).

One example of this trend has been the popularity of corporate or *strategic planning* (Lang, 1986b). Strategic planning seeks to cope with complexity by focusing on the development of strategies to guide organizational actions towards a particular objective. The approach is characterized by an action orientation, an emphasis upon implementation, a concentration on specific concerns, flexible and adaptive methods, and its attempt to build the learning capacity of participants (Lang, 1986b). Critics of strategic planning find fault on the basis that few organizations actually conduct strategic planning and that it only really provides a 'shell' within which conventional synoptic planning is implemented (Cayer and Weschler, 1988).

Another attempt to link strategic thought to planning implementation can be found in the work of Benveniste, who has forwarded the concept of *effective planning*, which he defined as being 'the management of change' (Benveniste, 1989: 264). Effective planning is both a political and a management activity. Planning is effective when it makes a difference. To make a difference, planning must invoke change. This implies that social power has been used. The basis of that social power is the political process and thus, effective planning is a political act (Benveniste, 1989).

According to Benveniste (1989), effective planning performs an important function within an organization. It provides a means of problem solving and it assumes a position of legitimacy, authority and power within an organization by providing a process that facilitates integration, consensus building and the management of uncertainty. Planning is based on the desire to handle uncertainty. He stressed the point that planning is embedded in organizations and that people within organizations seek to avoid making mistakes by controlling the environment within which they operate (Benveniste, 1981: 1989). Thus, planning is 'inevitably intertwined with management' (Benveniste, 1989: 127). The corollary is that for planning to contribute to organizational learning, planners and the planning function of an organization cannot be isolated from managers and the policy-making function of that organization.

Lastly, it was suggested that the scope of planning can be distinguished along three dimensions: comprehensiveness, the time horizon and levels of involvement (Benveniste, 1989). These criteria accord with the *hierarchy of planning* forwarded by Ozbekhan (1969, 1973). Ozbekhan's work built upon premises drawn from the field of management sciences to suggest a planning theory more adaptable to 'turbulent environments' characterized by interactive situations referred to as 'meta-problems' or 'wicked problems'. Planning may be viewed as the application of a range of methods to the design and attainment of desirable futures. Indeed, 'design is at the heart of planning' (Ozbekhan, 1973: 64) and is used to develop 'interventions' which direct action toward desired ends. The basis of this vision is the delineation of planning into three levels (Fig. 5.1; Smith, 1982a, 1984a).

In essence, the hierarchy of planning shown in Fig. 5.1 adapts the conventional approach to planning and places it within an explicit systems framework. In so doing, it seeks to enlarge and enrich the understanding of planning by providing for social learning as planning progresses from the normative though the strategic and to the operational level. This assumes that planning is conducted at each level and that there is a progression from more generalized policy making at the normative planning level, through the consideration of programmes at the strategic level, to the specifics of project implementation characteristic of operational level planning (Fig. 5.1).

In practice, opportunities for social learning in planning are often precluded. Frequently, normative and/or strategic planning only occurs after public opposition to the announcement of plans for project implementation at an operational level. Characteristically, this situation arises because the affected interests were only consulted at an operational planning level (Smith, 1982a, 1984a). In many instances, this is too late as key, framing decisions have already been made at the normative and strategic levels of planning. For example, a community might be asked for its views on the siting of a proposed waste management facility and its preference for site A or site B, a decision at an operational level of planning. The community has not been asked for its views regarding a policy for waste management (normative planning). Nor have they been consulted to determine a programme to consider preferences regarding which technology should be used and/or the appropriate criteria for locating a facility once the technology has been agreed upon (strategic planning). Thus, the public is asked to indicate a siting preference when it has not contributed to the prior planning responsible for that siting decision.

The notion of a planning hierarchy is useful as it reinforces the basic idea that planning should be concerned with the design of desirable futures and not just the immediate decisions necessary for project implementation. When that principle is applied to environmental planning, the goals of

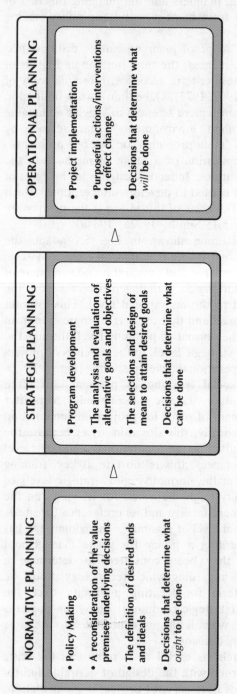

NORMATIVE PLANNING

- Policy Making
- A reconsideration of the value premises underlying decisions
- The definition of desired ends and ideals
- Decisions that determine what *ought* to be done

STRATEGIC PLANNING

- Program development
- The analysis and evaluation of alternative goals and objectives
- The selections and design of means to attain desired goals
- Decisions that determine what *can be* done

OPERATIONAL PLANNING

- Project implementation
- Purposeful actions/interventions to effect change
- Decisions that determine what *will be done*

Fig. 5.1 A hierarchy of planning
Source after Smith (1982a, 1984a)

planning are usually expressed in terms of improved decision making, the inclusion of environmental variables in planning, impact mitigation and conflict resolution. These same ideals can be expressed for impact assessment and there is a clear congruence between the planning process and impact assessment (Wood, 1988). However, the majority of impact assessment practice has been confined to the appraisal of project implementation at an operational level of planning. The challenge that remains is to determine how impact assessment should be manifest within the full hierarchy of planning.

Integrating impact assessment into planning

Thus far, three different approaches to the integration of impact assessment into planning have been undertaken. The first approach is the most conventional and it uses synoptic planning as the basis for a sequential model of impact assessment. The work of Westman (1985) is characteristic of this approach. A second approach involves the development of a manual as a guide to the application of impact assessment in planning. This approach is well represented by the manual developed for use in Britain by Brian Clark and others at the Centre for Environmental Management and Planning (Clark et al, 1981; Clark, 1983b). The third alternative to the integration of planning and impact assessment reflects a scientific approach to impact assessment, wherein the concepts and methods of social and ecological sciences are applied to impact assessment. A good illustration of this approach is found within the framework for impact assessment devised by Whitney and Maclaren (1985).

The synoptic approach

Figure 5.2 illustrates an example of the synoptic approach to the integration of impact assessment into planning. As outlined by Westman (1985), several 'phases' of impact assessment are based on an adaptation of the synoptic model of planning that involves:

- defining study goals
- identifying potential impacts
- predicting significant impacts
- evaluating significance of findings
- considering alternatives to the proposed action
- decision making
- post-impact monitoring.

As outlined by Westman (1985), the major elements of phase 1 involve the determination of data needs and resource requirements. Phase 2 focuses upon the boundaries of potential impacts, the range of those impacts and the estimation of significance. The prediction of significant impacts in phase

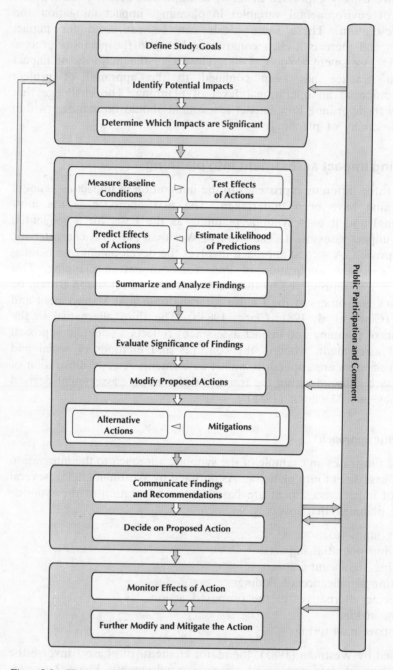

Fig. 5.2 The synoptic approach to the integration of impact assessment and planning
Source after Westman (1985)

3 rests upon the measurement of baseline conditions. These data provide the basis for projections and inferences regarding future conditions and the estimation of likelihood of impacts. The significance of these predictions is evaluated in phase 4 with the findings being summarized relative to the distribution of the impacts, how well the project goals are realized and the overall social significance of the impacts. In phase 5 the existence of alternatives to the proposed action is explored and steps that would mitigate the adverse effects of the proposal identified. This information is then available to guide decision making (phase 6) and post-impact monitoring (phase 7).

Within this approach, it is acknowledged that study goals are often predetermined (Westman, 1985: 10). Thus, although provisions for public input exist within phases 1 and 2, the impact assessment process largely involves the collection of data to meet predetermined objectives. Moreover, the data collection process itself tends to be reductionist. It is focused by the initial determination of significant effects and is heavily reliant upon the nature of the baseline measurements. By the time any alternatives to the project are considered (phase 5), a substantial investment of time and resources has already been made in studying the proposal. As a consequence, the project often becomes a *fait accompli*. Thus, the formal mechanism for decision making at phase 6 exists to ratify and legitimize acceptance of the project, with the promise of post-impact monitoring included as a final phase to appease concerns and placate any remaining opposition to the project.

Westman's model of the synoptic approach is fairly conventional and similar formulations can be found in Baldwin (1985) and Wathern (1988a). The synoptic approach represents the status quo and reflects the largely implicit manner with which impact assessment is viewed in planning. It is an approach that views impact assessment as a means to make decisions with ecological information in mind, without necessarily changing the basic outlook nor philosophy of that decision making.

The manual approach

In Britain, a manual has been developed to provide a guide to impact assessment in planning for use by government officials, developers and the public (Clark, 1983b; Clark et al, 1981). It seeks to specify both the requirements for information from developers and the criteria for project appraisal by government reviewers. The manual prescribes an approach to impact assessment practice involving:

• the acquisition of information
• the identification of likely impacts
• the appraisal of likely impacts
• the presentation of results
• public involvement and technical support.

Public participation is advocated throughout the various stages of the impact assessment process and is seen to be of specific utility in identifying impacts of public concern (Clark, 1983b). The approach to impact assessment advocated by the manual is shown in Fig. 5.3.

The initial step advocated in the manual is a form of pre-submission consultation with the developer to clarify and specify such items as siting criteria and the nature of the proposed development and its operations. A project specification report (PSR) is then prepared which provides initial

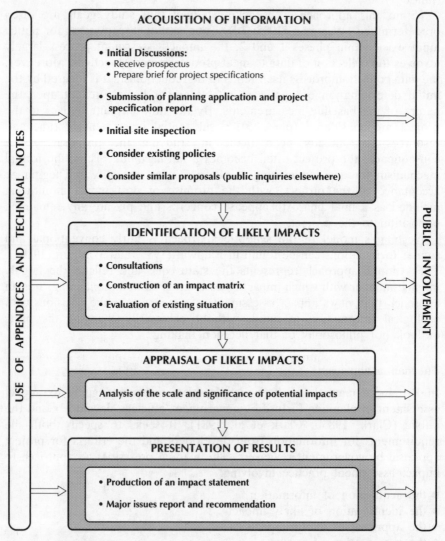

Fig. 5.3 The manual approach to the integration of impact assessment into planning
Source after Clark (1983b)

details regarding the physical characteristics of the proposed site; employment, financial and infrastructural requirements of the project; factors of environmental significance; emergency services; and any perceived hazards. This report provides the basis for the comprehensive appraisal of potential impacts.

Impacts are identified using an impact matrix which plots characteristics of the existing environment on one axis and activities in the construction and operational phases of the project on the other. An expanded matrix may then be employed to examine in further detail any specific impacts of known concern. Identified impacts can then be appraised, using an analysis of their scale and significance relative to whether the potential impact is beneficial or adverse, short term and/or long term, reversible or irreversible, direct and/ or indirect, and local and/or strategic. The findings are presented within an environmental impact statement (EIS) which details the project, its potential impacts and mitigating actions. The EIS should also contain a consideration of changes without the proposed project and a summary of the major issues suitable for dissemination as a separate report.

The manual presents a guide to the conduct of impact assessment. It suggests what data should be included in an assessment and how proponents should approach the evaluation of project impacts. Lastly, it provides a basis for government review of project proposals. Thus, it shares the strengths of other manuals in providing a clear checklist of the components that should be present in an impact assessment. However, manuals tend to be context-specific and are rarely portable from one set of environmental conditions, project characteristics and/or institutional arrangements to another.

The scientific approach

The scientific approach is predicated upon the view that the impact assessment process should be approached as if it were an experiment, subject to the rules and methodological stringency of science (Beanlands and Duinker, 1983; Whitney and Maclaren, 1985). Within this analogy, an existing environment is subject to disturbance resulting from the introduction of a project, process or policy. The impact assessor acts as the director of a scientific experiment in that (s)he attempts to predict the nature of the changes that will occur as a result of the development and determines their acceptability relative to previously defined criteria. If the development is approved, mitigative measures minimize negative consequences and the implementation of the development is monitored to measure the real (rather than the predicted) impacts. Further mitigative measures may then be invoked.

Clearly, the impact assessment process is distinct from a normal scientific experiment. The scale of the undertaking is usually far larger, the level of uncertainty far greater, the option to repeat the experiment does not exist and the distinction between 'good' and 'bad' effects is value-laden rather

87

than a precise measurement. However, despite these differences, the analogy to an experiment is useful as it implies the application of scientific rigour and high standards in the conduct of impact assessment studies: 'An *ecological approach* to environmental impact assessment is one that makes optimal use of ecological principles and concepts in the design and conduct of assessment studies and in the prediction of impacts' (Beanlands and Duinker, 1983: 18).

Using this definition, Beanlands and Duinker (1983) suggested an approach to impact assessment based upon the science of ecology. Key requirements outlined as the basis for an ecological approach to impact assessment were:

- the identification of valued ecosystem components
- the definition of a context for impact significance
- the establishment of boundaries
- the development and implementation of a study strategy
- the specification of the nature of predictions
- the implementation of monitoring.

The approach was seen as the basis for joint planning of an impact assessment between a proponent and an assessment review agency. The term *valued ecosystem component* was introduced to 'refer to the environmental attributes or components identified within the early, scoping phase of an EIA for which there is public and/or professional concern and to which the assessment should primarily be addressed' (Beanlands and Duinker, 1983: 18–19).

While Beanlands and Duinker (1983) articulated the rationale for an ecological approach to impact assessment, the concept of a scientific approach has been more fully expressed in a comprehensive framework for impact assessment devised by Whitney and Maclaren (1985). Their framework was developed around an ideal impact assessment process that involves scoping, prediction, significance assessment, evaluation, monitoring and mitigation, with public participation throughout (Fig. 5.4).

The scientific approach developed by Whitney and Maclaren incorporates an explicit social perspective to supplement the exclusively biophysical emphasis in the work of Beanlands and Duinker. In this manner, the Whitney and Maclaren model builds upon the concept of a scientific experiment and suggests that both social and ecological sciences are required elements of an integrated approach to impact assessment. Moreover, their approach was developed to counter perceived weaknesses in existing methodologies for impact assessment, including a failure to distinguish between the predictive and evaluative aspects of impact assessment, a tendency to focus on either the predictive or the evaluative aspect of methodology, and a failure to understand the interplay between the objective and subjective aspects of impact assessment methodology (Whitney and Maclaren, 1985).

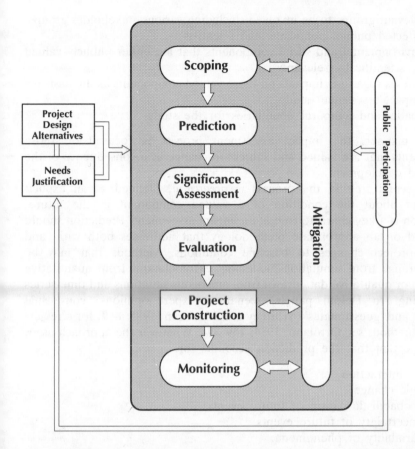

Fig. 5.4 The scientific approach to the integration of impact assessment into planning
Source after Whitney and Maclaren (1985)

The Whitney and Maclaren framework outlines an ideal process for impact assessment that is initiated in response to a project development. Two steps that precede the impact assessment process itself are the determination of *project design and alternatives* and the attention given to *needs justification*. In determining the engineering feasibility and economic viability of a project, it is desirable that a proponent consider various alternatives to the undertaking from the outset. It is equally as important that the proponent has given adequate justification for the proposed project and that the question of need has been clearly established. *Public participation* may assist in these determinations and the framework provides for public input at all phases of the impact assessment process (Fig. 5.4).

Following these antecedents, the first step in the assessment process is *scoping*. Scoping provides a preliminary scrutiny of the proposal and it requires the determination of:

- the relevant groups to be involved (including various stakeholder groups, the affected public, politicians and scientists)
- the environmental and social components that are either publicly valued and/or scientifically relevant
- hypotheses regarding impacts on those valued components as the basis for the impact assessment study
- the spatial and temporal boundaries for the study.

In this manner, the impact assessment process is focused on those components that are valued and subject to change as a consequence of the proposed development.

The second step in the process is *prediction*. Defined as an explicit statement about the condition of a valued component in the future, prediction is a key characteristic of an impact assessment. Prediction should be based on an appropriate methodology that addresses both with- and without-project changes to baseline conditions. Methods that may be utilized range from simulation modelling, extrapolations from quantitative data and/or qualitative data drawn from analogous situations and time-series forecasting, to Delphi panels, scenarios, expert opinions, consensual methods and 'guesstimates' (Whitney and Maclaren, 1985: 8–9; for a review of such methods see Ortolano, 1984: 159–77). Whatever the approach used, it is important that the prediction accounts for:

- system interaction
- dynamic change
- the probabilistic nature of future events
- the uncertainty of future events
- the variability of phenomena.

Having predicted the changes that will affect the various valued components, it is then necessary to determine their significance. Thus, within *significance assessment*, the magnitude of the changes forecast for each component is assessed to determine its significance. Impact assessment methods have often not explicitly considered the issue of significance (Thompson, 1990). Thus, it is important that the criteria used to determine significance include:

- without-project comparisons
- cumulative effects
- impact duration
- risk
- the stability and resilience of the environmental components
- relative magnitude
- quality standards.

Evaluation refers to the process whereby the positive and negative changes in all the components are considered. On the basis of that determination, decisions are made regarding whether or not to proceed with the development and, if so, which alternative(s) should be adopted. A variety of tools has been developed to aid the evaluative process, including dominance analysis, weighted summations and the use of utility functions (see Whitney and Maclaren, 1985: 13–19; Ortolano, 1984: 181–212; Smith and Theberge, 1987). These tools have emerged as attempts to address the following issues that arise in evaluation:

- aggregation versus disaggregation of impacts
- the level of measurement employed
- weighting procedures
- the use of common measurement units
- ease of understanding
- the number of alternatives to be evaluated
- public involvement.

The impact assessment process does not end with the implementation of development. If, and when, a development is approved its actual effects must be monitored. Three kinds of *monitoring* can be identified:

- *compliance* or *surveillance monitoring* to ascertain concurrence with specified mitigation measures and guidelines during project construction and operation
- *effects monitoring* to compare actual with expected impacts and to validate predictions
- *public concerns monitoring* to identify and track the views and opinions of the public affected by the construction and operation of an undertaking.

Relatively little has been written about impact monitoring techniques. It has been noted that the practice of impact monitoring can be improved through the use of audits (Bisset and Tomlinson, 1983; Munro et al,1986; Bailey and Hobbs, 1990), but most writers have simply paid lip service to the need for monitoring. There have been few attempts to specify how impact monitoring should take place or who should have responsibility for conducting and assessing the monitoring programme once it is in place. One exception was the study by Krawetz et al (1987), who outlined a framework for monitoring in impact assessment involving a monitoring plan, process management and a monitoring objective. They utilized this framework to examine three case studies in Canada and found that the management process played a key role in effective monitoring but that satisfying impact management and prediction objectives at the same time was extremely difficult.

The final element of the Whitney and Maclaren framework is that of *mitigation*, defined as the actions undertaken to ameliorate the negative impacts of a development. As shown in Fig. 5.4, mitigation measures may be recommended at the scoping stage, at the significance assessment stage and during monitoring. Mitigation may involve a number of actions, including:

- changes in design, in processes, of raw materials and/or in location
- remedial measures
- the use of visual barriers to hide developments
- compensation, both in monetary and non-monetary terms.

Again, as with monitoring, there have been relatively few studies that have looked at the compliance record of proponents with mitigation requirements. However, one study in California has indicated that the enforcement of mitigation measures is highly variable and that weaknesses in monitoring contribute to poor compliance (Johnston and McCartney, 1991).

As described by Whitney and Maclaren (1985) the scientific approach offers a framework for the integration of impact assessment in planning that is a normative ideal. This is both its greatest strength and its most apparent weakness. As an ideal, the framework not only identifies the elements necessary for impact assessment but establishes the imperative for those elements to be approached with a clear study design. In turn, that design should be bounded by scientific principles and conform to the standards and rigour of a scientific experiment.

Clearly, this is an ideal and the process outlined may not be feasible nor applicable in its entirety in all situations. The framework can also be criticized for ignoring the fiscal, resource and time constraints that can be imposed on impact assessment studies. However, those constraints should not be accepted as immutable. Moreover, neither the synoptic nor the manual approaches to the integration of impact assessment in planning offer the same potential for critical appraisal as the scientific approach and, of the three alternatives, it provides the best framework from which to assess current practice and determine future needs.

Current practice

Based on reviews of the Canadian situation (Beanlands and Duinker, 1983; Smith et al, 1989; Smith, 1989), it is possible to identify some clear distinctions between the normative ideal of the scientific approach and the descriptive reality of current practice in impact assessment:

- The question of need is rarely determined explicitly in impact assessment studies. Rather, the merit of a proposed development is usually assumed or justified in general and unspecified terms such as the 'greater economic good'.
- Environmental and social effects are rarely significant criteria during the initial design phase of developments and alternatives to the undertaking are usually limited in number and restricted in their scope.
- The element of prediction is often absent. Instead, impact assessments are based upon extended descriptions of baseline environmental conditions and a consultant's view of pending changes.
- Significance assessment is usually subsumed within the evaluation

component of an impact assessment. It is at this phase that poor science is most in evidence in impact assessment practice.

- Mitigation and monitoring are often viewed as synonymous entities, routinely being covered with a blanket statement that 'actual impacts will be monitored and mitigated as necessary'.

Thus, rather than the ideal process shown in Fig. 5.4, assessment practice is often closer to an exercise of environmental inventory, 'expert' comparison and proposed mitigation. These steps provide for environmental assessment but they lack the prediction, significance assessment and evaluation necessary for *impact assessment*. This contrast highlights the need for reform in how impact assessment is integrated into planning.

Looking at both the Canadian and the international experience, Marshall et al (1985) discussed the 'growing consensus on general principles for reform' in impact assessment processes and procedures. The need for reform stems from several interrelated and existing weaknesses in impact assessment practice. Generally, impact assessment lacks a clear policy framework. The absence of a suitable context for policy planning leads to isolated reviews that promote *ad hoc* planning and result in impact assessment becoming a 'surrogate vehicle for public debate' regarding project justification (Marshall et al, 1985: 8). Overlaps and omissions in institutional arrangements lead to lengthy reviews, extended approvals procedures and proponent frustration. In turn, these give rise to residual questions of due process, with the equity and efficiency of impact assessment open to criticism. Exacerbating the problem are the science-based deficiencies of impact assessment and its perceived inability to deal with uncertainty, especially at the prediction, significance assessment and evaluation phases of an impact study. Lastly, a further weakness results from a general lack of monitoring and post-project audits and the failure to follow through and follow up on impact assessment studies (Marshall et al, 1985).

To counter these weaknesses in impact assessment practice, Marshall et al (1985: 12) advocated the structural reorganization of decision-making processes to develop the 'appropriate planning prefix and implementation suffix' for impact assessment. Central to this reorganization would be a more effective integration of impact assessment with project planning and the resource management framework under which it operates.

This call for reform and the implementation of impact assessment at all levels of planning is not new. For example, Lee and Wood (1978) advocated a hierarchical or 'tiered' approach to impact assessment, while Sewell and O'Riordan (1981) addressed the need for both policy review and project appraisal through impact assessment. However, it is important that impact assessment not be viewed as an adjunct to the planning process but, rather, as a process for planning in and of itself. What is needed is a redefinition of impact assessment, one that moves impact assessment away from its project implementation, impact statement, methodological focus and towards a wider view of impact assessment as a planning process to achieve the goals of sustainable development.

CHAPTER 6

Impact assessment redefined

Introduction

Impact assessment must be redefined. The rationale for this reform was introduced in Chapter 1, as was the imperative for resource management to adopt sustainability as a guiding philosophy and principal goal. A conceptual framework is needed that can serve as the basis for initiating appropriate environmental planning and policy for sustainability. In Chapter 1, it was suggested that impact assessment originated with a desire to be consistent with these goals but that it has failed to fulfil them.

The basis for this failure was examined in Chapter 2, which reviewed the state of the art in impact assessment methods and methodology. It was argued that appropriate methods exist to conduct impact assessment but that the science of impact assessment has not been strong. Moreover, the dominant paradigm for impact assessment has emphasized the production of an impact statement as the means to provide information to decision makers regarding the appraisal of project implementation. This is a narrow focus and it has inhibited the ability of impact assessment to address such issues as risk, uncertainty and cumulative effects. Presently, impact assessment has good technique but poor process.

The need is for impact assessment to become integral to environmental planning and resource decision making rather than just serve as a check upon them. This is a new role for impact assessment. It requires the redefinition of impact assessment and the reconsideration of its function within a broader resource management context. To this end, impact assessment should be redefined as *a process of environmental planning that provides a basis for resource management to achieve the goal of sustainability.* This definition requires that impact assessment become a bridge to integrate the science of environmental analysis with the politics of resource management. Central to this design are the institutional arrangements for decision making that define the provisions for impact assessment in resource management (Ch. 3). These provisions must then be understood and implemented within the framework of components established by Chapters 4 and 5, namely: the policy process, interest representation and impact assessment as a process of environmental planning.

This chapter outlines a framework for sustainable resource management. The framework provides the means by which the various elements affecting impact assessment may be integrated. As such, it provides the basic frame of reference for understanding the newly defined role of impact assessment relative to the goal of sustainability and the broader context of resource management. The chapter concludes by outlining the data sources used and procedures employed in the application of the framework to the international experience with impact assessment in the three case study chapters that follow.

Sustainable resource management

There are three basic components to sustainable resource management. First, problems must be identified. Second, proposals for policies, strategies and projects to respond to perceived problems must be derived through a process of resource management. Third, sustainability exists as a desired end point for resource management in the resolution of problems.

Problem identification

The starting point for resource management should be the identification of problems. Within any particular locale, problems may arise in the economy, in society and/or in the environment. Often problems involve complex interrelationships among all three. In each instance, the identification of problems varies as a function of several interrelated variables including problem tractability, the justification of need, issue attention and information availability.

The notion of problem *tractability* was introduced in Chapter 3. It recognizes that some problems are more amenable to solutions than others. Problems are easier to manage (more tractable) when they are well understood and a lot is known about them, they affect fewer people and they are the result of predictable actions or behaviours. Conversely, problems become harder to resolve (intractable) when understanding is lacking and little is known about them, they affect larger numbers in society or if there is a high degree of uncertainty about the causes and effects of activities. Thus, the problems surrounding the location of a sewage-treatment plant are inherently more tractable than those involved in resolving the question of global warming.

Tractability influences problem identification in a number of ways. Both individuals and organizations tend to focus their efforts on problems that appear to be amenable to solution. In contrast, issues that appear to be more intractable are often avoided. A direct corollary is that pressing concerns that are complex, involve major behavioural changes and/or about which there is much uncertainty, can be passed over in favour of problems that may be more trivial in nature but are open to easy or more immediate solutions.

Need is a relative term. In resource management, need is often invoked to help justify a preferred course of action. For example, a proponent might argue that a specific policy is needed to rectify a problem, or that a particular project is warranted as the most appropriate solution to an issue. For example, a dam might be proposed as the solution to a perceived water-shortage problem. Under these circumstances, a predetermined solution is justified on the basis that it is 'needed' to resolve the problem at hand. However, the problem itself has been defined from the perspective of the proposed solution. The dam will increase water supply and the problem has been defined in terms of a shortage of available water supply. In fact, the perceived water shortages may indeed be temporary owing to abnormally low rainfall in a given year. In these circumstances, water conservation measures may suffice to reduce demand below the critical level, and would negate the need for a costly dam to be constructed.

As in this example, the existence of a preferred course of action often acts to narrow the definition of the problem and tends to restrict the range of issues considered. The ends become defined in terms of the means by which they may be attained. The means may then be justified on the basis that they are 'needed' to meet those desired ends. It is critical therefore that any examination of problem identification consider the justification of need and the extent to which real needs have been determined.

The identification of problems will also be greatly affected by the degree of attention generated by an issue. The more attention generated, the more likely it is that the full range of problems will be identified. *Issue attention* varies with the type and extent of media coverage commanded by (or accorded to) an issue, the numbers and nature of the public affected and the perceived priority on political agendas. Downs (1972: 38) suggested that 'public attention rarely remains sharply focused upon any one domestic issue for very long' and developed the notion of an 'issue-attention cycle' to characterize the systematic progression of changing public attention to issues. According to Downs, issues go through a cycle in which they attain sudden public prominence, remain there while the public's enthusiasm and interest remain high and then gradually recede into the background as the economic and social costs of resolution are realized and/or the issue is found to be intractable. Although the issue may remain unresolved, it fades from the centre of public attention. Concomitantly, public attention has already focused on another issue (Downs, 1972).

The last factor affecting the way in which problems are defined is *information availability*. Information on resource issues can be derived from a number of sources. The data may be biophysical, ecological, social, economic, political and/or geographical. The availability of these data, and the ability of science to provide precise measurements of various phenomena, can have a significant bearing on the determination of present conditions, the understanding of causal relationships and estimates of the future. These are central elements in how problems are defined. A key question remains. Does the information provide the necessary knowledge for problems to be identified fully and correctly?

Often, the existing information provides an adequate description of conditions, allowing problems to be identified, but it does not provide a functional understanding of the situation. Dorcey (1986) discussed this distinction well. He used the example of the Fraser River in British Columbia to illustrate the differences between *descriptive knowledge* (such as the distribution of juvenile salmon and the location of marshes) and *functional knowledge* (such as how spawning salmon are affected by decreasing marsh habitat). The distinction extends to the ways by which new knowledge is obtained. New descriptive knowledge is derived from inventories and monitoring. In contrast, new functional knowledge can arise only out of hypothesis testing and analysis (Dorcey, 1986). Thus, for problems to be identified correctly it is preferable that both the descriptive information be sufficient and the functional understanding of those data be adequate.

Taken in conjunction, problem tractability, need justification, issue attention and information availability influence how problems are defined. The rationale for resource management and its specific purpose at any given point, are derived from problem identification. Remaining unresolved are the questions of who should then manage the resource, how and for what purposes?

Resource management

According to Mitchell (1989: 3), resource management 'represents the actual decisions concerning policy or practice regarding how resources are allocated and under what conditions or arrangements resources may be developed'. Resource management has evolved greatly over the past three decades. Once considered to be a purely technical exercise, resource management has been subjected to tremendous reform pressures. Throughout the 1970s and 1980s resource managers adjusted to a variety of demands calling for a more pluralistic, more comprehensive and more integrative approach to resource management. As a consequence, it is often the *process of resource management* that is of paramount concern to agencies, managers and the public. The key variables that affect this process are the institutional arrangements for management, interest representation and impact assessment as a process of environmental planning.

Institutional arrangements provide the structure for resource management.

Effective planning and decision making are facilitated when appropriate institutional arrangements are in place. The effectiveness of those arrangements can be modified at a number of leverage points encompassing:

- the *context* within which management occurs
- the *legitimation* of relevant management authority, responsibilities and jurisdictions
- the management *functions* being assigned
- the *administrative structures* used to provide for management
- the *processes and mechanisms* used in management
- the *culture and attitudes* of participants.

Resource management should be viewed as a political process that determines the direction of, and the control over, the manner by which resources are developed. This process tends to be administered through organizations and agencies that can greatly influence the manner by which policies are both formulated and implemented. Administrative organizations are decision making entities. Decision processes represent a means for resources to be allocated and conflicts to be resolved. Decision making itself arises from a bargaining process that should involve the various affected interests. Thus, a central issue that arises within resource management is the determination of the interests that require representation within decision-making processes.

Interest representation seeks to ensure that all of the relevant stakeholders are afforded an opportunity to participate in decision making. It is a recognition of the need to give people input into the decisions that affect them. The term stakeholders is used to refer to all the parties who have a vested and legitimate interest in participating in the resource management process. Typically, the stakeholders in a resource management issue would include a project proponent, management agencies, various pressure groups, the lay public affected by a proposal and other selected interests. Their interests could encompass a wide spectrum of goals and objectives, and would be expressed through the various means of interest representation in resource management such as lobbying, public participation and/or environmental dispute resolution.

Interest representation is necessary within the policy process to ensure political commitment to the implementation of resource management practices such as impact assessment. It also provides stakeholders with an opportunity to influence future planning efforts. A central concern of planning is the design of desirable futures and the effective management of change. Planning can occur at a normative level (with a focus upon policies to guide what ought to be done), at a strategic level (with an emphasis upon the determination of programmes to guide what can be done) and at an operational level (with a concern for project implementation and what will be done in practice). At each level, *environmental planning* should invoke the principles of *impact assessment* in seeking to provide for improved decision making, impact mitigation and conflict resolution. Indeed, impact assessment should be viewed as a process for environmental planning.

Thus far, however, impact assessment has been integrated into the process of planning but has not been viewed in a broad enough perspective to be seen as a means to achieve management objectives. If impact assessment is to be regarded as a process for environmental planning, the notion of impact assessment solely for project appraisal has to be discarded. This requires that the role of impact assessment within planning be clarified and that the practice of impact assessment recognize the importance of the following activities:

- scoping
- prediction
- significance assessment
- evaluation
- monitoring
- mitigation.

Sustainability

The concept of sustainability was discussed in Chapter 1. While it remains difficult to define precisely, sustainability is a term that is both useful and easily understood as a desired goal for resource management. Aspiring to sustainability directs resource management efforts towards a balance among societal, economic and environmental objectives. Development, and not just growth, becomes paramount. Human desires are balanced by the necessity to preserve environmental integrity, while pressures for economic growth are balanced by the requirements for societal equity. The purpose of resource management is to supply the basis by which this balance may be achieved.

As a goal for resource management, sustainability represents a vision of a desirable future. It is a goal that can guide both the design of future plans and the more immediate management of present changes in society, its economy and the environment. Global awareness of environmental issues has never been higher. Translating that awareness into concerted political action requires an easily recognized icon around which to build support for more effective policies and the implementation of integrated resource management practices. Sustainability provides that symbolism.

An integrative framework for sustainable resource management

The various components of problem identification, resource management and sustainability can be linked through an integrative framework for sustainable resource management. As shown in Fig. 6.1, the framework proceeds from a starting-point of problem identification towards the goal of sustainability, with resource management supplying the bridge linking them together. The framework defines a role for resource management procedures and institutions. It involves the translation of values and information into directives for sustainability. Within this paradigm, impact assessment is seen as the process for environmental planning that provides the basis for resource management to achieve the goal of sustainability.

Problems in a society, its economy and/or the environment are perceived on the basis of issue tractability, the justification of need, issue attention and the availability of information. As issues are identified, the process of resource management is initiated to determine what actions (if any) should be undertaken to resolve the situation. Once the resource management process has been initiated, it influences problem perception. Thus, a feedback loop exists between resource management and problem identification, allowing for revisions in the determination of problems as the resource management process progresses through various iterations.

Resource management represents the central component of the framework. As shown in Fig. 6.1, resource management is envisaged as an iterative process, wherein three elements must be integrated:

- institutional arrangements
- interest representation
- impact assessment.

As discussed in Chapter 3, *institutional arrangements* must be viewed in context and may be assessed on the basis of legitimation, functions, administrative structures, processes and mechanisms, and culture and attitudes (Fig. 3.1). Each of these factors represents a potential leverage point for which opportunities exist to alter the institutional arrangements for resource management. Within the integrative framework shown in Fig. 6.1, the institutional arrangements are viewed as structuring the resource management process. They determine the role to be played by interest representation and they establish the provisions for impact assessment (Fig. 6.1). Thus, to be effective, the institutional arrangements must provide an appropriate basis for the integration of interest representation with the process of impact assessment.

Interest representation is shown in Fig. 6.1 as a function of the stakeholders involved, their goals and objectives, and the approaches used to make representation within decision making. As discussed in Chapter 4, three principal approaches to interest representation exist: lobbying, public participation and environmental dispute resolution. Typically, these approaches are employed by proponents or management agencies to involve the input of other stakeholders such as pressure groups, lay citizens and selected interests within the resource management process. In any resource management issue, it is important that the full range of stakeholders be

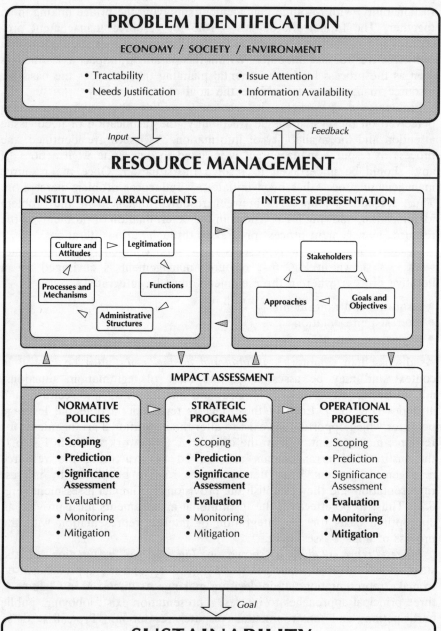

Fig. 6.1 An integrative framework for sustainable resource management

identified and approaches to interest representation be adopted that are appropriate to the various goals and objectives of the different stakeholders. For this reason, Fig. 6.1 deliberately fails to show any linear relationships between stakeholders and specific approaches to interest representation. This approach reinforces the notion that it is more pertinent to ensure the representation of all stakeholders within the resource management process than it is to adhere to a distinct method of involvement. Which stakeholders are involved, their views and how they are represented may change several times through the various iterations of the management process.

The final component of the integrative framework is that of *impact assessment*. As shown in Fig. 6.1, impact assessment is a process for environmental planning. This view of impact assessment amalgamates the concept of a hierarchy of planning (Fig. 5.1) with the scientific approach to impact assessment (Fig. 5.4) outlined in Chapter 5. Planning can occur at three different levels, involving the design of normative policies, strategic programmes and/or operational projects. At each level, six activities are seen as necessary: scoping, prediction, significance assessment, evaluation, monitoring and mitigation. Where planning occurs at all three levels, the iterative sequencing of these activities would involve a shifting emphasis with normative planning stressing scoping, prediction and significance assessment; strategic planning emphasizing prediction, significance assessment and evaluation; and the final operational planning focused around the activities of evaluation, monitoring and mitigation (Fig. 6.1).

Thus, impact assessment is seen as the engine within the resource management process. Impact assessment operates within the provisions provided by the institutional arrangements and it must incorporate the demands of interest representation. However, it is impact assessment as a *process* that provides the basis for resource management to achieve the goal of sustainability. In this manner, the redefined role of impact assessment is given full expression. Central to impact assessment are the stages of prediction, significance assessment and evaluation. Impact assessment is distinguished as a design-centred process of environmental planning by an explicit attention to these three activities.

Prediction is the activity that distinguishes impact assessment as a planning process as distinct from a descriptive assessment of existing environmental conditions. Through prediction, impact assessment is concerned with the future effects of proposals on society, its economy and the environment. Prediction addresses the nature of change, both with and without the implementation of the proposals under scrutiny, and establishes the basis for the significance of those impacts to be assessed and evaluated. In contrast, an environmental assessment that simply provides an analysis of existing conditions can only be used as the basis for inference regarding proposed changes. In the absence of prediction, there can be no explicit attention to, or forecasting of, future conditions as the basis for an assessment of impact.

Significance assessment and *evaluation* follow logically after prediction.

Once future conditions have been forecast, the impacts of proposals must be assessed to determine their relative significance and magnitude. Significance assessment should consider such factors as cumulative effects, impact duration, and the stability and resilience of the systems being affected. The relative positive and negative changes can then be evaluated. Evaluation represents the stage in impact assessment where trade-offs must be determined, the various alternatives weighed against one another and the choice of a preferred course of action resolved.

For impact assessment to function as a process of environmental planning, the three activities of prediction, significance assessment and evaluation must all be present. Moreover, while they are activities that are closely related, it is important that each be present as a discrete activity within the assessment process if a systematic and thorough consideration of future impacts is to be derived.

Within the framework, the proponent is identified as one stakeholder in the resource management process. In many circumstances, the proponent assumes a far greater role, often having responsibility for initiating and commissioning an impact assessment for development proposals it wishes to have approved. Indeed, one reason that other impact assessment models have been project-specific is because this role is so prevalent. In contrast, the present framework seeks to be more generic, addressing not only project-specific decisions, but also policy development and strategic planning initiatives in resource management. Thus, as shown in Fig. 6.1, the integrated framework allows for any lead actor to be specified from the range of stakeholders having an interest in a particular proposal.

Lastly, the framework does not identify a single decision path by which the three components (institutional arrangements, interest representation and impact assessment) may be integrated. Rather, a series of differing interactions is conceived as being both possible and potentially effective. The principal characteristic of the framework is that it recognizes that the three components must interact if resource management is to provide an effective bridge between the basic starting point of problem identification and the desired goal of sustainability.

Reviewing the international experience with impact assessment

The integrative framework for sustainable resource management is used in the following three chapters as the basis for an evaluation of the international experience with impact assessment. The intent of this review is to:

- illustrate the range of current practice in resource management relative to the role of impact assessment and the goal of sustainability
- illustrate the utility of the framework as both a normative model for a

redefined role for impact assessment and as an evaluative schema in the analysis of resource management practice.

The framework is applied in the three case study chapters using data derived principally from available, published literature. An attempt has been made to focus the discussion on well-known examples that would be illustrative of both the range of current practice and of the utility of the framework. On this basis, general comments are made about the experience with impact assessment in frontier developments (Ch. 7), for linear developments (Ch. 8) and within waste management (Ch. 9). The chapters offer substantive illustrations of impact assessment under conditions of differing institutional arrangements. The examples used are international in scope and were selected to highlight the variability within which the various key concepts developed over the first six chapters of the book have been put into operation.

The examples demonstrate the relative strengths and weaknesses of existing approaches to impact assessment relative to the 'new role' as defined within this chapter, using the following structure:

- general outline and problems
  - technology involved
  - design issues
  - concerns to be addressed by impact assessment
- illustrative cases
  - general literature overview
  - selected studies in depth
- indications from the application of the framework regarding the current status of impact assessment practice.

CHAPTER 7

Frontier developments

Introduction

Frontier developments involve large-scale projects for resource exploitation. They are often in remote locations, where conditions are technically very challenging. They are characterized by advanced technologies, high capitalization and a high degree of uncertainty. Frequently, frontier developments are promoted on the basis that they will provide jobs in areas where unemployment can be very high. However, many frontier developments have also caused severe impacts on aboriginal peoples.

For most environmentalists, frontier developments are undesirable. They involve large-scale projects such as big dams, big power stations or big mines that are, as a genre, perceived to be environmentally unpalatable (O'Riordan, 1990). Big projects are viewed as harmful to the environment, invariably inducing environmental stress and community distress. In frontier regions, these impacts are borne disproportionately by local (often aboriginal) groups who 'usually have to suffer in silent protest, or who must leave their lands to an even more vulnerable existence as ecological refugees' (O'Riordan, 1990: 141). Moreover, it has become increasingly evident that the economic benefits of the projects themselves are often being impaired by environmental stresses that shorten the life expectancy of resource schemes and generate secondary side-effects that are expensive to remedy. For example, premature sedimentation, seepage and excessive evaporation have been frequent problems seriously affecting the economics of large dams, especially in the tropics (Goldsmith and Hildyard, 1984).

Some of the difficulties of predicting the impacts of frontier developments were examined by Berkes (1988). Using the James Bay hydro megaproject in Canada as his frame of reference, Berkes demonstrated that the accuracy of most impact predictions is low. The notion of impact assessment considering all impacts associated with a project is, therefore, a fallacy. Indeed, in the James Bay instance, most impacts were either predicted incorrectly or were not predicted at all. While poor science contributed to these deficiencies, two factors were seen as major barriers to improved impact assessment: (1) the continuing lack of knowledge concerning large hydroelectric projects, and (2) the intrinsic unpredictability of impacts (Berkes, 1988).

These findings underscore the importance of uncertainty (regarding the technology, its implementation and its likely effects on the environment) and of scale (frontier developments are simply too big to conform to conventional linear planning concepts) with respect to impact assessment for frontier developments. In response, O'Riordan (1990: 141–2) has suggested that 'indices of change in local community well-being' should also be incorporated into the planning of large-scale projects, with frontier developments being viewed as 'packages of environmental and social services' rather than as isolated projects.

Several factors define the character of planning for frontier developments. First, projects tend to be promoted on the basis of larger, supraregional benefits usually expressed in terms such as national or regional development goals. At issue for an impact assessment is the extent to which the question of need for the specific project has been examined: is the project warranted by the larger goals or is the project itself being used to help justify these ulterior objectives?

Second, benefits from frontier developments usually accrue to distant urban populations but impact on local, rural and/or aboriginal populations. This factor gives rise to a potential clash of cultures and the imposition of cultural values on minority populations. Thus, issues that arise include determining how well questions of social impact are integrated into the overall process of impact assessment. Is the process of impact assessment being used as a tool for integrative planning or is the response one of *ad hoc* planning? A key concern at this juncture is the nature of the institutional arrangements for management and the absence or presence of a suitable policy environment in support of a more holistic approach to planning.

Third, frontier developments are characterized by a high degree of uncertainty. The scale of project and its possible impacts are very large, the impacts of technology unclear, and base data often lacking or inadequate. For impact assessment this reinforces the importance of adequate scoping, the determination of valued environmental components and the scientific credibility of predictions and significance assessments.

In conjunction, these factors emphasize various design issues and, at a normative planning level, often generate an *ad hoc* planning response. They provide the basic *context* for impact assessment of frontier developments, and emphasize the need for that impact assessment to focus on the initial phases of scoping, prediction and significance assessment. Thus, the key questions that must be addressed in the case of frontier developments are:

● *Project justification*:
· What are the perceived benefits of the project?

- · Has the question of need been explored?
- • *The terms of reference for impact assessment*:
  - · Is environment defined broadly to include both socio-cultural and biophysical elements?
  - · Will planning provide for integrated management of resources?
  - · What are the existing institutional arrangements for management?
  - · Are there existing policies to guide project development?
  - · How are the relevant stakeholders to be defined and represented in the planning process?
- • *Uncertainty*:
  - · What technology is involved?
  - · What are the associated risks?
  - · What is the status of base data for the region?
- • *Impact assessment science*:
  - · How are the issues to be scoped?
  - · What methods of prediction are to be used?
  - · How is significance assessment to be determined?

This chapter illustrates how impact assessment has been applied to frontier developments. Selected experiences with impact assessment are reviewed using the examples of oil and gas in Canada's Arctic, hydro megaprojects and regional development projects. The integrative framework developed in Chapter 6 is then used to illustrate the potential for change in the practice of impact assessment for frontier developments.

Oil and gas in Canada's Arctic

The discovery in 1969 of significant hydrocarbon reserves in northern Alaska led to the extensive exploration for, and subsequent promotion of, hydrocarbon production in Canada's Arctic (Fig. 7.1). Awareness of the potential impact of this development was heightened after proposals for a natural gas pipeline through the Mackenzie Valley prompted the establishment in 1974 of a public inquiry chaired by Justice Thomas Berger (Sewell, 1981). After a three-year examination and a cost in excess of $5.3 million, Berger issued his final report, characterizing the North as both a homeland for its small resident population and as a resource frontier for the distant majority of Canadians (Berger, 1977). In dramatizing this distinction, the Berger Inquiry acted to raise public awareness of Northern issues, stimulated political development in the North, and secured a 10-year moratorium on hydrocarbon production in the region (Smith, 1987a).

The Berger Inquiry established a number of precedents for the conduct of impact assessment in Canada, particularly in regards to interest representation (Smith, 1987a, 1990c). Among its procedural innovations were:

- • a broad interpretation of the inquiry's mandate, which required that the potential impacts of the proposed development be assessed from both an environmental and a social perspective

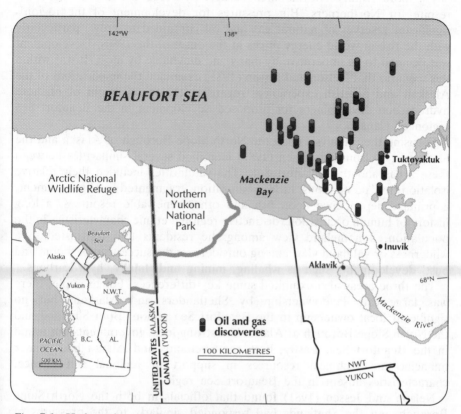

Fig. 7.1 Hydrocarbon exploration in the Beaufort Sea region

- the use of three different hearing formats: preliminary hearings for a rudimentary scoping of the issues; formal hearings for technical debate and the presentation of the proposals; and a series of community hearings throughout the North where local residents could voice their concerns
- the provision of funds to allow a range of native, environmental and local interest groups to participate throughout the inquiry
- the acceptance of individual experiences and concerns as valid information on par with technical, scientific and other 'objective' types of information
- extensive use of the media to publicize, inform and gain support for the inquiry and its findings.

In so doing, the Berger Inquiry established the concept of social impact assessment in Canada and provided a normative model for the practice of public participation in resource management (Smith, 1987a). The success of the Berger experience also inspired belief in Canada in an emerging approach to resource decision making: impact assessment.

The moratorium on development secured by Berger provided a window of respite for Northerners. But pressures for development of the region's significant reserves of natural gas and oil continued to grow, particularly with the rise in world energy prices at the onset of the 1980s. Development was deemed to be imperative by many and inevitable by most. It was within this scenario that Nelson and Jessen (1981) examined the implications of the Alaskan and Scottish experiences regarding the management of offshore hydrocarbon development for proposed development in the Beaufort Sea region of Canada.

Focusing their analysis upon the North Slope Borough of Alaska and the Shetlands, Nelson and Jessen (1981) identified several similarities between these areas and the Beaufort Sea. The similarities included their relative isolation, low populations, little urbanization, fragmented local government, a primary dependence upon fish and other renewable resources, a long history of human occupation, distinctive regional ethnic characteristics in the population, a homeland view among the residents that contrasted to a wilderness or wildland view among outsiders, and a succession of 'boom and bust' developments such as whaling, mining and, finally, hydrocarbons.

The three areas also exhibited some key differences. For example, there was large-scale landownership by Shetlanders and Alaskans but no significant local ownership in the Beaufort Sea region. Both Shetland and the North Slope Borough of Alaska had strong local governments not found in the Beaufort Sea. Lastly, both the Shetlands and Alaska had greater financial and technical resources in support of regional self-reliance, characteristics absent in the Beaufort Sea region.

Nelson and Jessen (1981) found that officials in both the North Slope Borough and the Shetlands had responded similarly to the pressures of petroleum development. Both had made major changes to the institutional arrangements for management of the development and both had established procedures for more co-ordinated planning. There was a strong interest in improved monitoring of environmental conditions in both places but neither had adequate arrangements for inspection nor surveillance during the construction and operational stages of development.

Based on the Alaskan and Scottish experiences, Nelson and Jessen (1981) advocated that several measures be undertaken in the Beaufort Sea:

- strengthen and improve local government's ability to manage development
- improve the project approval system
- provide land for native stewardship and economic activities
- modify the institutional arrangements for shore zone planning to provide for better regional control over agency co-ordination and the environmental data base and better interest representation.

Nelson and Jessen's study pre-dated a formal proposal for development of production facilities in the Beaufort Sea region. However, a proposal for hydrocarbon production was made in 1980 and was referred to a formal review under the federal environmental assessment review process (EARP). Previously, EARP had been applied to proposals for other hydrocarbon projects in the North but without much success (King and Nelson, 1983). The Beaufort Sea environmental assessment review process (BEARP) represented an important step in the evolution of resource management in Canada's North as it 'involved a significant expansion in the use of EARP by the federal government for planning and management of northern development' (Sadler, 1990: 3). The final report of the BEARP was filed in July 1984, after an extensive and lengthy public inquiry costing $2.7 million and a total impact assessment bill of over $20 million.

The Beaufort Sea environmental assessment review process

The BEARP was convened to assess the concept of hydrocarbon development in Canada's western Arctic, an area with significant natural gas and oil reserves (Fig. 7.1). The companies involved (Dome, Gulf and Esso) submitted a nine-volume EIS outlining the industry's thoughts about the probable interactions between the environment of the Beaufort Sea region and the possible scale, pace and siting of potential production facilities. The potential scale and impact of any proposed development were enormous and could have committed the region to a 'path of industrialization' that would have shaped 'the course of development across the entire North' (Sadler, 1990: 3). Thus, the sub-text of the review concerned the potential for adaptation of indigenous cultures and subsistence economies to the apparent imperatives of a frontier development.

In its final report, the BEARP review panel concluded that oil and gas production would be socially and environmentally acceptable if appropriate measures were undertaken. These measures involved commencing production on a small scale, with only gradual expansion; construction of a small-diameter, buried pipeline to transport natural gas southwards; the potential use of tankers to transport oil; and the assumptions that monitoring and surveillance mechanisms to be in place prior to project approval and mitigative measures would be applied. Thus, the BEARP final report gave a limited endorsement to the industry to proceed with its plans but failed to resolve any of the uncertainty surrounding the important process and management issues that would affect the implementation of those plans.

In reviewing these recommendations, the Canadian Arctic Resources Committee (CARC, 1984: 2) contrasted BEARP to the Berger Inquiry, noting that both had a 'broad mandate to deal with fundamental issues of northern development and to make recommendations that would reach far beyond the fate of a specific development project'. The Berger Inquiry fulfilled this mandate and established a standard against which other impact assessment efforts will always be measured. In contrast, the BEARP was

111

more costly, took longer to complete and was considered to have failed in its mandate (CARC, 1984).

The failings of the BEARP were partially circumstantial. Changes in world oil prices, disappointing exploration results and spiralling financial difficulties combined to diminish industry pressures for development projects in the Beaufort Sea. (Indeed, by 1986 when viable oil reserves were finally proven in the Beaufort Sea, the impetus for development was almost non-existent.) However, Rees (1984) suggested that the principal source of the BEARP's failings was its inability to utilize its broad terms of reference to define the institutional arrangements necessary to manage development proposals in the future. Rather, the BEARP created high expectations, seemingly adopted procedural innovations in support of those expectations but then conspicuously avoided addressing the fundamental issues needed to achieve them. For those seeking to establish impact assessment as 'a rigorous and useful part of decision-making', the BEARP report was 'a profound disappointment and setback' (Fenge, 1984: 16).

The federal government commissioned its own evaluation of the BEARP experience (Sadler, 1990) which sought to determine if the application of EARP in the Beaufort Sea instance had provided for impact assessment with:

• rigorous technical analysis
• responsive consultative procedures
• responsible institutional arrangements
• relevant decision-making implementation.

The evaluation indicated that the BEARP adopted a conventional approach to impact assessment of attaching terms and conditions for project approval and management through the production of an EIS (Sadler, 1990). However, the scope and complexity of the review panel's mandate were such that a conventional approach was largely unable to focus the review on significant issues, despite some noteworthy procedural innovations (such as an issues seminar which attempted to provide scoping for the assessment). In the Beaufort Sea example, a regional development proposal combining policy, programme and project-specific elements required assessment. The conventional EIS-based approach proved inadequate to the task as too many generalizations and assumptions had to be made about possible project designs, technologies and mitigation capabilities in the absence of firm project specifications. Moreover, in focusing its efforts on providing a more technical EIS, the panel failed to consider fully the policy requirements and institutional arrangements required for the management of developments in the future.

Hydro megaprojects

In the 1970s, large scale resource development schemes (megaprojects) emerged as a dominant trend in resource management. Often, the central focus of these schemes involved energy development, especially hydroelectric facilities. As Sewell (1987: 498) noted, the philosophy of hydro mega-projects 'appealed to enterprising businessmen and engineers as a bold and imaginative means of getting to the future faster'. By definition, almost all hydro megaprojects involve the construction of a large dam. Big dams are promoted on the basis that they will be of great economic and social benefit. Politically, they usually have great appeal as they offer an opportunity to use economies of scale and stimulate regional economic expansion. Big dams are seen as sources of cheap, clean power for industrialization. They can prevent damaging floods and can provide irrigation for agricultural development.

However, large dams have been plagued by many problems (Goldsmith and Hildyard, 1984). Among the criticisms levelled at large dams are claims that they:

- cause massive dislocation of local populations, cultures and sustainable economies
- fail to provide promised economic benefits, especially in terms of agricultural development
- suffer from environmental difficulties, including threats to entire ecosystems, which generate more problems than existed prior to dam construction
- often impinge upon the rights of native peoples
- provide minimal benefits to impacted, local populations, while providing much larger benefits to sponsoring agencies, construction companies and distant urban populations.

Some of the best-known examples of hydro megaprojects are those in Africa. Food production has declined steadily in Africa over the past 20 years. Serious instances of environmental degradation have accompanied this decline. Thus, the objective of most hydro megaprojects has been to harness water resources better to provide for more food production (through such techniques for water resource management as irrigation and flood protection) and to contribute to development through hydroelectric power production and improved transportation (Le Marquand, 1989).

Usually, these projects have entailed the construction of dams to store water and regulate its use for power generation, urban water supply and irrigation. For multi-purpose projects, large dams were justified on the basis that 'the hydropower generation component provided a highly profitable return that subsidized other economically marginal project components' (Le Marquand, 1989: 128). To illustrate this point, Le Marquand used the example of the Manantali Dam on the Senegal River. A 'classic multipurpose project', the dam was designed to:

- provide for a regulated downstream flow of 300 m^3 s^{-1}
- permit navigation 900 km upstream
- produce 800 GWh of electrical energy per year
- have 100 MW of guaranteed power

- reduce the 100-year peak flood to that of the 10-year peak
- enable double-crop irrigation of 255 000 ha.

It remains to be seen whether the dam can fulfil these projections, or if it will become another example of a hydro megaproject failing to perform to its design expectations, such as the Aswan High Dam or the Kariba Dam. Economic returns have often been disappointing and unforeseen environmental impacts have caused such harm to dependent ecological and social systems that river basin developments 'rather than being a solution to the challenges of economic growth and development had become, on balance, a part of the problem' (Le Marquand, 1989: 130). Moreover, the focus on power generation 'has led to a neglect of other development opportunities' in agriculture, fisheries, forestry and transportation (Le Marquand, 1989: 131).

Yet, the development of hydro megaprojects in Africa must be placed in the proper context wherein development is subject to external pressures (including declining terms of trade for agricultural commodities, the world debt crisis and declining financial flows into Africa) and internal constraints which include uneven resource distribution, rising populations and environmental stresses (Le Marquand, 1989). Reporting upon a major study of the African experience with river basin development, Scudder (1989: 139–140) noted that a narrow focus on river basins as hydrological systems 'has tended to restrict development, on the one hand, to water resource management, and, on the other hand, to national accounting in terms of electricity generation and crop production on large-scale irrigation projects'. A perennial failure has been the tendency for river basin development to be equated with the supply-side management of water resources through the use of structural alternatives. This bias has been further enforced by institutional arrangements that confer the responsibility for river basin development to agencies mandated to provide for energy, irrigation and public works projects rather than planning. The result has been an overestimation of the benefits to be derived from hydropower generation and irrigation and an underestimation of the environmental and socio-economic costs of dams. Thus, the study's most important single conclusion was that 'African nations have been developing the hydroelectric potential of their rivers at the expense of their ecological resiliency, human populations and agricultural potential. Furthermore . . . the provision of cheap hydropower has not stimulated development to the extent expected' (Scudder, 1989: 140–1).

The African experience has been mirrored elsewhere (Goldsmith and Hildyard, 1984; Goodland, 1990b). But despite this mixed track record, the worldwide popularity of hydro megaprojects continues seemingly unabated. In Canada, for example, the dominant test of impact assessment in the 1990s will be the response to the Great Whale hydro megaproject proposed for construction in Quebec's James Bay region. Following previous extensive hydro developments in the region, the Great Whale megaproject would cause substantial environmental impacts and it poses a serious threat to the viability of the indigenous Cree culture. Elsewhere, other controversial hydro megaprojects proposed for the 1990s include the Pak Mun Dam in Thailand (which threatens the homes of over 20 000 people, farmland, tropical rain forest and a National Park), the Three Gorges project in China (which will require the relocation of over 1.4 million people) and a series of dams in Brazilian Amazônia, such as the Samuel Dam and the Ji-Paraná Dam (which threaten vast tracts of tropical rain forest and Indian lands).

However, the economic recession of the early 1980s, 'lower demand, low or zero economic growth, combined with crippling debt burdens and evaluations' did force a re-evaluation of capital-intensive big hydro projects (Goodland, 1990b: 110). Hence, the construction of some hydro mega-projects has been postponed, while others have been cancelled. For example, the San Roque and Binongan projects in the Philippines were both cancelled after delays induced by opposition to their completion (Abracosa and Ortolano, 1987), while the Nam Choan project in Thailand created sufficient public concern over potential losses of tropical rain forest to lead to its cancellation (Phantumvanit and Nandhabiwat, 1989). Meanwhile, in such countries as the United States, Canada, Australia and New Zealand, awareness of the potential for massive environmental and social disruption from hydro megaprojects led to several major conflicts in the 1980s between project proponents and environmental groups opposed to their construction (Sewell, 1987).

Using proposals for the Franklin Dam in Tasmania, the Clyde Dam in New Zealand and the Site C Dam in British Columbia as illustrative case studies, Sewell (1987: 500–1) identified that hydro megaprojects are favoured 'when responsibilities are in the hands of a large public utility which is firmly supported by the government, and where the prevailing economic philosophy is that of "hydro-industrialization"'. The key deter-minants of a hydro megaproject were considered to be the institutional arrangements for management and the influence of the various stakeholders within them (Sewell, 1987). Electric power utilities were shown to have played a major role in all three case studies. Finally, Sewell (1987) concluded that both the role of the utilities and the philosophy of hydro-industrialization should be the focus of increased scrutiny in the future.

An example of attempts to reform the institutional arrangements for energy management was examined by Smith (1988a). Also using the Site C hydro megaproject as an illustrative case, Smith's study centred upon the reform of energy planning in British Columbia invoked by the passage of the 1980 Utilities Commission Act. For the first time, the province's electric utility (BC Hydro) was subject to regulation. The proposal for the Site C Dam was filed with the newly created BC Utilities Commission only two weeks after its enabling legislation was approved. Thus, it represented the first instance where the new regulatory regime could be applied to a BC Hydro project (Smith, 1988a).

The Site C project

The Site C proposal involved the construction of a dam and a 949 MW hydroelectric generating station on the Peace River, 7 km south-east of Fort St John in British Columbia (Fig. 7.2). Budgeted at $3.5 million, the Site C project faced opposition on the basis that it was not needed and that the reservoir it would create would inundate a substantial amount of agricultural land. Whereas the latter fear had more emotional appeal, it was the former concern which led to the eventual postponement of the project.

BC Hydro's proposal was reviewed by a five-member panel established by the newly created BC Utilities Commission (Smith, 1988a). The Site C review panel adopted a public inquiry format modelled after that used by the Berger Inquiry and concluded that the necessary certification for the project 'be withheld and deferred until the essential evidence of need and project justification be demonstrated' by BC Hydro (Smith, 1988a: 436).

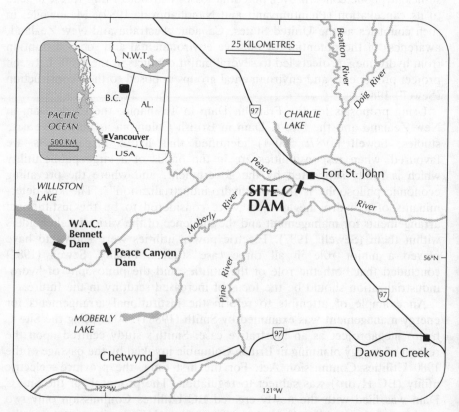

Fig. 7.2 Location of the proposed Site C Dam in British Columbia

The crux of the Site C review had been the consideration of need for the Site C Dam to be built. During the public review of the proposal, it became clear that there was insufficient existing demand for electricity to warrant Site C being built and that the utility's determination of project justification had been in error (Smith, 1988a).

The Site C decision marked the first time that a major project of BC Hydro had not been given routine endorsement and approval. The new Utilities Commission Act had altered the ground rules for project approval in British Columbia. A more thorough and open system for energy planning and the assessment of project impacts had been created. Concomitantly, the autonomy of the provincial electric utility, BC Hydro, over project decision making had been removed (Smith, 1988a).

The Site C review acted as a catalyst for change within BC Hydro. New management was appointed and the utility was severely downsized during a three-year transition away from an energy-project development focus. Smith (1988a) contended that this change and its ramifications resulted from the revised institutional arrangements for energy planning constituted by the 1980 Utilities Commission Act. In effect, the new legislation had acted to 'tame' BC Hydro and initiate a new approach to energy development within the province.

Regional development projects

Increasingly, hydroelectric power generation is but one component of a larger regional development project. For example, the Tucurui Dam in Brazil was built to support the development of mining projects as part of the Greater Carajás programme in the Amazon. Indeed, the exploitation of Amazonia in Brazil is perhaps the best known recent example of a frontier regional development project.

Legal Amazonia covers over 5 million km^2 or 60 per cent of Brazil (Fig. 7.3). Estimates suggest that anywhere from 7 to 12 per cent of this area has been deforested since 1970 as a consequence of logging, clearing for ranching, mining activities, and the construction of roads, military bases and hydro megaprojects (Goldemberg and Durham, 1990; Fearnside, 1990; Goodman and Hall, 1990b; Hall, 1989): 'During the last 20 years, the Amazon has undergone a process of accelerated transformation, of which destruction of the forest is simultaneously an instrument and a symptom' (Goldemberg and Durham, 1990: 22). International concern over deforestation in Amazonia has centred around concerns over global warming. Indeed, as concern has grown, many people in temperate countries have begun (incorrectly) to associate destruction of the Amazonian forest with the fundamental cause of climatic changes (Goldemberg and Durham, 1990). In fact, global changes in climate are largely a function of the developed world's dependency upon the automobile and the production of greenhouse gases throughout the northern hemisphere.

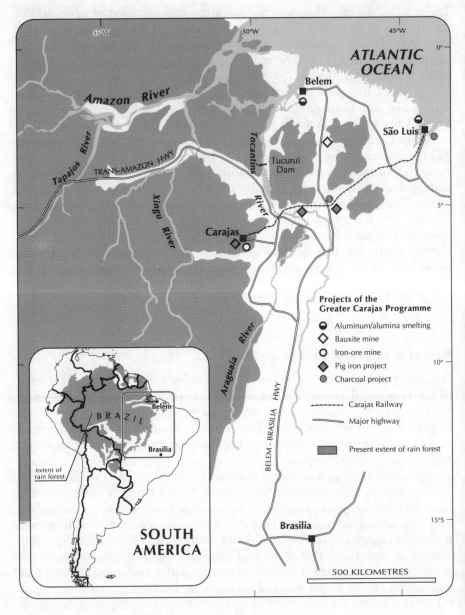

Fig. 7.3 Frontier development in Amazonia

Whatever its impetus, this concern has led to a questioning of the regional development strategy followed in Amazonia. Prior to 1970, development in the Amazon region had been mostly confined to the early rubber boom of 1890–1910 and the establishment in the 1950s of a series of large cattle ranches. In 1970 the Brazilian government moved to promote a concerted policy of regional development for Amazonia. Within the twin goals of national integration and defence, the development push was based upon an extensive programme of road building, colonization and agribusiness (Hall, 1989; Goldemberg and Durham, 1990; Hecht and Cockburn, 1990). By the 1980s, mining had replaced cattle ranching as the region's major investment priority and regional development in Amazonia has since been best characterized by the vast Greater Carajás development (Hall, 1989; Goodman and Hall, 1990b).

Estimated at a cost of over $62 billion, the Greater Carajás programme encompasses a quarter of the Amazon Basin (or nearly 11 per cent of Brazil). It combines an iron-ore mine, two integrated aluminium plants, a 900 km railway, pig-iron smelting and steel production, hydro megaprojects (including the Tucurui Dam) and an associated array of infrastructural, agrolivestock and lumbering activities (Goodman and Hall, 1990b). At the heart of the development is a huge iron-ore mine, which will utilize the world's largest known reserves to supply upwards of 7 per cent of the world supply of iron-ore per annum (Hall, 1989; Marquès, 1990; Neto, 1990). The development has been managed by Brazil's biggest iron-ore producer, Comphania Vale Rio Doce (CVRD), with considerable financial support from many of the world's largest iron-ore importers and a variety of international banks and agencies, including the EC Development Fund, the World Bank and Eximbank of Japan (Marquès, 1990; Goodman and Hall, 1990b; Neto, 1990).

The Carajás project has been the source of considerable controversy, spurred by concerns for the ecological and social damage its components would incur and by the contribution of international organizations to the viability of the development. Among the most severe impacts of the Carajás development are the loss of over 2400 km^2 of land and the displacement of Indian bands due to the reservoir created by the Tucurui Dam and the potentially huge losses of rain forest involved with the plan to locate a series of pig-iron smelters alongside the project's rail link (Hall, 1989; Fearnside, 1990). The development has also escalated the displacement of local Indians and peasants, prompted further environmental degradation and contributed to the growing problem of rural violence in the region (Goodman and Hall, 1990a). Yet, as Goldemberg and Durham (1990, 35) summarized, the conditions imposed on the development by such agencies as the World Bank have proven to be insufficient

> because they encompassed measures only to combat the negative effects. They did not change the priorities to show that development in and of itself was not the primary objective, except when allied with preservation of the tropical forest, its self-sustained productive management and non-predatory forms of occupying the territory.

In other words, the World Bank and other lending institutions have been negligent for failing to insist on *sustainability* and not just economic growth as the principal criterion for support of regional development projects. These concerns highlight the role of international organizations in providing funds for regional development projects and their role regarding the implementation of impact assessment in developing countries (Horberry, 1985).

The role played by the World Bank

The International Bank for Reconstruction and Development (the World Bank) was established in 1945 as part of the United Nations system. Its major function is to provide loans for development projects and it is the largest of the international development banks (Jéquier and Weiss, 1984a). In 1988, the Bank invested in 217 projects in 83 countries, with project loans in fiscal year 1988 totalling $19.2 billion (Goodland, 1990a).

In the process of fulfilling its mandate as a development agency, it has been suggested that the World Bank has become a major technological institution. That is, it has become an organization that 'finances or carries out research, transfers technology, promotes better choices and uses of technology, encourages innovation, and contributes to the building up of technological capabilities' (Jéquier and Weiss, 1984a: 1). Thus, the World Bank does more than simply provide financial support. It seeks to achieve socio-economic objectives in support of development. To fulfil this role, the Bank functions as a technological institution, actively promoting certain technologies and the mobilization of local resources to implement those technologies. Indeed, the Bank's technological function may be viewed as 'an inevitable, and indeed necessary, component of its activities as a financing institution and development agency in the conception and execution of projects and in the formulation of sectoral development strategies' (Jéquier and Weiss, 1984b: 326). The Bank's promotion of technology may be integral to its function, but does it help or hinder the achievement of sustainability?

Traditionally, the World Bank financed development projects emphasizing capital infrastructure (especially roads, railways, ports and power generation facilities) and basic industries (such as agriculture, forestry and mining). From the 1970s onwards the Bank began to diversify and provide loans focusing upon development projects aimed at alleviating poverty by meeting basic needs and redistributing income (including projects in the education, health, water supply, sewerage and population sectors). In conjunction with this attention to social concerns, the Bank also began to incorporate environmental criteria into its appraisal of development projects.

An Office of Environmental Affairs was established at the Bank in 1970. However, over the next 15 years the environmental record of the World Bank was frequently challenged and in the 1980s the position of the Bank was strongly attacked for 'its failure to mitigate the environmental destructiveness of many development projects . . . [and for] . . . supporting virtual ecological and human catastrophes' (Le Prestre, 1989: 33). Foremost in these criticisms was the World Bank's support of the Greater Carajás programme (Goodman and Hall, 1990b; Goodland, 1990a).

Le Prestre (1989) examined the evolution of the Bank's environmental policy during this era. His figures indicate that 37 per cent of funded World Bank projects in the period 1971–78 'had immediate negative environmental consequences' (Le Prestre, 1989: 29). Emphasizing the organizational constraints it faced, Le Prestre contended that the Bank was restricted in its ability to play a more effective role in the protection of natural resources in developing countries.

For example, project appraisal is the sole responsibility of the Bank. But rather than transforming its administrative structure, the Bank responded to the need for reform by adding an Office of Environmental Affairs. The Bank rejected the use of mandatory impact assessments on the basis of:

- cost
- a lack of staff to conduct reviews
- the perceived generality of impact statements
- delays in project implementation that would be incurred
- the absence of appropriate environmental policy frameworks in developing countries (Le Prestre, 1989: 55).

Rather, it aimed to quantify environmental concerns and to modify its principal tool for decision making, benefit–cost analysis. However, as a financial institution the Bank was staffed largely by economists. Thus, the Bank was inclined to tackle environmental problems from an economic perspective which assumed that growth enables societies to solve environmental problems, that environmental problems are quantifiable and that the market mechanism and individual property rights are sufficient to resolve most problems (Le Prestre, 1989: 61).

Le Prestre (1989: 197) concluded that the performance of the World Bank was constrained by the number of contradictions it faced:

- between its development policies and the environment
- between the training of its staff and the need for an environmental perspective
- between incentives and environmental values
- between its goals as a banker, developer of infrastructural projects and demands for environmental protection.

In 1987 the Bank moved to implement a new policy placing environmental protection at the centre of its work (Le Prestre, 1989). While this represented a 'profound organizational reform of the Bank's operations', few donor countries or loan recipients had pushed strongly for this shift in priorities (Le Prestre, 1989: 188–99).

The nature of the reforms and the implementation of the new environmental policy by the World Bank have been detailed by Goodland (1990a, b). The Bank's objective in introducing environmental policies into its practices was to ensure that, with careful planning, adverse effects 'can be prevented, mitigated, or compensated for and the beneficial effects enhanced' (Goodland, 1990b: 112). Projects eliciting World Bank support must be designed by a member government and expect to earn a profit (normally, the internal rate of return must exceed 10 per cent). They are then eligible for loans ranging from 30 to 70 per cent of total project costs. The World Bank utilizes a project cycle for reviews that provides for:

- environmental reconnaissance
- environmental assessment
- appraisal
- supervision
- completion and post-audit.

Usually, the Bank and the proponent government will have co-operated prior to the submission in a pre-identification phase. This reconnaissance provides a preliminary screening of projects and initiates the assistance of environmental specialists as integral members of the project design team (Goodland, 1990b). The Bank prefers to integrate environmental concerns into the project design so that by the time a 'project reaches feasibility, most of the major environmental risks will have been designed out' (Goodland, 1990a: 150). Furthermore, the Bank avoids adding an impact statement to a previously designed project on the rationale that any redesign would be wasteful, that impact statements are 'a recipe for confrontation', and that 'add-ons are the first to be dropped in times of budgetary cutbacks' (Goodland, 1990a: 151).

At the appraisal phase, the Bank stresses the need for institutional strengthening and training. Major projects are supported by a 'triple alliance' that involves: (1) an in-house environmental unit within the project ministry, (2) a strengthened national environmental ministry or its equivalent, and (3) an independent review panel of experts. Project supervision is provided by Bank staff, guided by specific policies and operational directives. Finally, the completed project is audited and further post-audits initiated when necessary (Goodland, 1990a, b).

The World Bank's approach is thorough and as Goodland (1990a: 152) has suggested, if all the various procedures and policies were 'conscientiously applied, projects would have few, if any, environmental problems'. However, this has not been the case, and projects have had problems. Goodland (1990a: 152) identified the Greater Carajás programme as a classic example of a 'superb project amidst a sea of environmental abuse'. Such failings result from economic development being narrowly viewed simply as the product of a collection of projects. Thus, while the project approach is flexible, it has limits and the focus must shift to national environmental quality and sustainable resource management (Goodland, 1990a). To meet these challenges, the Bank has begun integrating environment into the economic dialogue of member governments, has started to operationalize sustainability, and is attempting to revise the economic calculus applied to the determination of externalities (e.g. Daly, 1990).

However, the real test of the World Bank's revised sense of environmentalism rests with the projects it rejects and approves. In recent years the Bank has withdrawn its support or declined to finance several large hydropower developments on environmental grounds (Goodland, 1990b). But critics of the World Bank remain sceptical about its ability to deliver on its promises of environmental integrity and give little credence to such responsible actions. Instead, they focus on the Bank's continued involvement in such controversial and high-profile frontier developments as the proposed flood control plan for Bangladesh (Boyce, 1990; Brammer, 1990).

Flood controls in Bangladesh

In 1989, the World Bank prepared *An Action Plan for Flood Control in Bangladesh*. This report was in response to a request arising from the 1989 G-7 Summit Meeting in Paris for the Bank to co-ordinate international efforts to alleviate the effects of flooding in Bangladesh in the wake of the devastating floods of 1987 and 1988 (Brammer, 1990; Boyce, 1990).

Flooding is an annual event in Bangladesh (Boyce, 1990). The Bengal Delta and the associated floodplains of the Brahmaputra, Ganges and other rivers comprise over 80 per cent of the country (Fig. 7.4). In the summer monsoon season as much as 83 per cent of the floodplain can be inundated by floodwaters. These annual floods are beneficial and are termed *barsha* in Bengali (Boyce, 1990). They fertilize the soils of the region, provide the basis for rice cultivation and promote fish harvesting. In contrast, floods that are abnormally deep or persistent are referred to as *bonna*. These floods represent a significant threat. In 1988, the flooding was particularly severe with several million people forced to abandon their homes and over two-thirds of the capital city of Dhaka flooded (Boyce, 1990).

The World Bank advocated a five-year agenda for flood control involving pilot programmes, flood protection for Dhaka and further studies aimed at the preparation of a flood master plan, at an estimated total cost of $146.3 million (Brammer, 1990; Boyce, 1990). The idea for a flood master plan had been originally proposed by a policy options study funded by the United Nations Development Programme (one of four studies that preceded the World Bank report: the others were two engineering proposals commissioned separately by the French and Japanese governments and a US-funded study of flood mitigation options). The plan forwarded by the World Bank was an uneasy compromise among these competing proposals and it contained an apparent endorsement of the ambitious French scheme calling

123

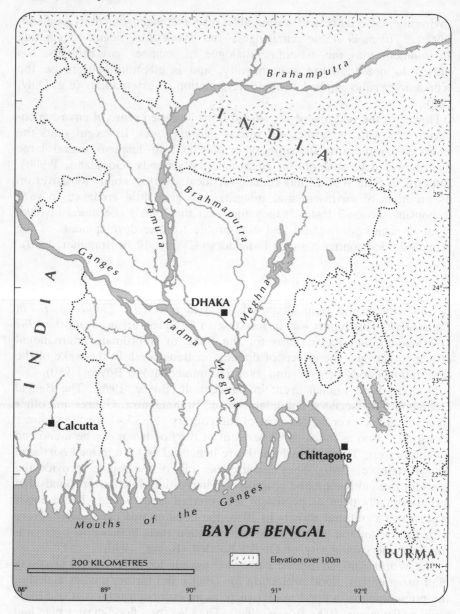

Fig. 7.4 Flood vulnerability in Bangladesh

for embankments and major structural solutions to the flood problem. The French approach is favoured by the military government of Bangladesh and it calls for high embankments (4.5–7.4 m high) to be built along all of Bangladesh's major rivers at a total construction cost estimated in the range of $5–$10 billion, with operating costs of $160–$180 million per year (Boyce,

1990; Brammer, 1990). It is this aspect of the plan, and the Bank's apparent endorsement of it, that has caused concern.

Both Boyce (1990) and Brammer (1990) identify a number of critical issues arising from the embankment scheme. These include:

- The environmental risks involved, which threaten the ecological balance of the region and the fisheries in particular. The risks stem from a general uncertainty associated with megaprojects and a more specific lack of basic information (e.g. on the effects of flooding on soil fertility in the region).
- Need: the 1988 flood was exceptional (1 in 100) and the proposal would not necessarily prevent its repeat but would stop the annual (beneficial) floods, which are not perceived to be hazardous.
- A concern that the embankment scheme attempts flood prevention rather than flood protection. Many experts question the feasibility of channelling such big, meandering rivers as the Brahmaputra and the Ganges, which can be up to 20 km wide, have sediment loads of over 2 billion tonnes and can experience lateral channel movements as high as 800 m per year. This task is especially challenging in a region prone to major earthquakes.
- Claims that the economic feasibility of the proposal has been overestimated by over-inflating the potential benefits that the project might bring to Bangladesh.

Boyce (1990) also referred to an internal World Bank audit of other flood control projects in Bangladesh. That report cited failures arising from a lack of institutional mechanisms for implementation of flood management measures and other problems, such as the construction of housing on embankments using fill necessary to the hydraulic integrity of the structures. Thus, there is an added sense of unease stemming from the Bank's lack of adequate scrutiny of previous projects in Bangladesh. This concern is reinforced by indications that the Bank has allocated insufficient funds to allow a proper impact assessment of the project (Boyce, 1990; Brammer, 1990).

Critics fear that the World Bank action plan advances an agenda which predetermines adoption of the French scheme. In essence, this view challenges the notion of policy change at the World Bank and suggests that far from adopting a new approach, the Bank is still operating under the premise of 'business as usual'.

Redefining impact assessment for frontier developments

At the onset of the chapter, the basic characteristics of frontier developments were identified. Consequently, impact assessment for frontier developments must address the particular issues posed by project justification, the terms of reference for impact assessment, uncertainty, and impact assessment science. The four examples outlined above (Beaufort Sea oil and gas, the Site C hydro megaproject and regional developments in Amazonia and Bangladesh) illustrate the manner by which impact assessment has been applied to frontier developments. How successful have those efforts been, and what reforms are needed to improve the practice of impact assessment for frontier developments? Answers to these questions can be derived through a review of the four examples using the framework described in Chapter 6.

Problem identification

Tractability

Although all frontier developments have elements that are difficult to resolve, the issue of tractability is best illustrated by the problems inherent in attempts to control flooding in Bangladesh. The situation is complex, with both societal and economic motivations for a swift and effective response. In Bangladesh it is simply not possible to modify public response to the hazard potential through translocation. The population is too large, the hazard zone is too extensive, and alternative locations are already overcrowded. Thus, plans have focused on controlling the flood hazard through some variation of structural responses. However, potential engineering solutions are complicated by a number of factors, including:

- the meandering nature of the rivers that form the Ganges Delta
- the size of the rivers themselves
- the risk of earthquakes
- the requirement that the threat to life posed by the *bonna* be curtailed but that the life-sustaining *barsha* continue.

It is not clear that the flood hazard can be successfully engineered at all. Yet, the task facing impact assessment is to determine the ramifications of any proposed engineering solutions on the social, economic and environmental sustainability of the region. In any circumstances this would be a difficult task, but, for Bangladesh, it is rendered all the more intractable because the region is agriculturally dependent, overpopulated and impoverished. Thus, while the problem itself is life-threatening, possible engineering solutions run the risk of being just as devastating.

Needs justification

Many frontier projects are proposed under the guise of 'regional economic development' and their supposed financial benefits are taken at face value, with only minimal and/or superficial scrutiny. A presumption of technical proficiency is usually extended to project proponents and it is assumed that they have correctly determined the economic viability of their proposal. The number of large projects that have failed to fulfil their economic expectations is testimony to the fallacy of this proposition.

The Site C hydro megaproject provides an excellent illustration of the benefits to be gained from questioning a project's fiscal integrity and the economic rationale for its construction. Without a review of needs justification under the BC Utilities Commission, it is probable that the citizens of British Columbia would have inherited a large and expensive lake of limited ecological or recreational value rather than a productive generating station had the proposed dam at Site C been built. The review emphasized the absence of any need for immediate construction. The proponent's failure to adequately justify project need directly contributed to a moratorium on its construction.

In contrast, the need for the Greater Carajás programme was assumed without any extensive exploration or justification. That project exemplifies the notion that economic development in a frontier region has intrinsic merit: that the benefits of industrialization and resource exploitation are imperative and of sufficient inherent value to outweigh any costs involved. In such circumstances, the project is approved first and the only task open to impact assessment is the determination of the terms and conditions for its implementation.

Issue attention

The Beaufort Sea environmental assessment review process (BEARP) illustrated very well the influence of issue attention on impact assessment. In the years immediately prior to the BEARP, oil and gas exploration had been a growth industry in the Beaufort Sea. Rising world prices favoured expansion and a rapid movement towards production. In anticipation of forthcoming proposals, the BEARP was established with high expectations and with much fanfare. It was to determine the terms and conditions under which oil and gas from the Beaufort Sea were to be produced and transported. However, prices did not continue to rise and actually declined. Industry plans were postponed and no specific proposals existed for review. The issues were imprecise and clouded with uncertainties. Public attention focused elsewhere. The BEARP report was released four years later to a reception that was passive at best. Time and circumstances had simply passed the review by.

Information availability

A lack of information can prevent present conditions from being accurately described. It prohibits causal relationships from being understood and inhibits the ability to make forecasts. In short, a lack of information negates the basis for impact assessment and undermines the integrity of planning.

Frontier developments

The development of Amazonia illustrates this phenomenon. The vastness of the region and its ecological complexity have compelled exploration and fostered many grandiose dreams. However, the Amazon remains inadequately understood. The current loci of development are proceeding without adequate knowledge of baseline conditions or an understanding of dynamic ecological interrelationships. The result is a haphazard mixture of megaprojects, infrastructural expansion and enclosure whose cumulative effect threatens catastrophe. Many of these projects have undergone 'planning' but the results stand as testimony to the axiom that planning without information is as good as no planning at all. This is not to suggest that full knowledge should become a precondition for development. Rather it is to point out that at present nearly all developments in frontier areas start before enough information has been collated to permit sound planning through effective impact assessment.

Summary

In conjunction, these factors indicate a number of difficulties for the impact assessment of frontier developments at the problem identification stage:

- Sustainability has not yet been identified as a central issue in problem formulation. Most projects originate as solutions to economic concerns, with the environmental effects of proposals receiving acknowledgement but not full design integration. By and large, social aspects are not formally integrated into the identification of problems but are supposedly subsumed within economic criteria.
- Engineering feasibility still predominates as a design criterion. Moreover, impact assessment is not being utilized to assist in the determination of design feasibility.
- Impact assessment faces a challenge to be timely and integrative but not time-consuming. In frontier areas, with limited data bases and large expanses of terrain, it is a particularly difficult challenge.

Institutional arrangements

The examples reveal that the role of institutional arrangements for impact assessment is complex and multi-faceted. However, despite their apparent differences, the institutional arrangements for impact assessment of frontier developments are consistent with the following scenarios:

- well-developed institutional arrangements where impact assessment acts as the initiator of reforms e.g. the Site C hydro megaproject
- well-developed institutional arrangements where impact assessment is the target of reforms, e.g. Beaufort Sea oil and gas

- situations where the institutional arrangements are poorly developed and there is a reliance upon international aid agencies to foster the implementation of impact assessment, such as the Amazon and Bangladesh.

Impact assessment as the initiator of reform

The Site C impact assessment review demonstrated that the institutional arrangements for energy planning in British Columbia had been altered by the passage of the 1980 Utilities Commission Act. The hydro megaproject acted as a catalyst for that reform. Changes at one leverage point within the previous institutional arrangements led to ramifications and adjustments in the remaining leverage points as follows:

- *Legitimation*: passage of the Utilities Commission Act indicated the government's political commitment to reform the province's approach to energy planning and its provisions for impact assessment. Administrative responsibility for the implementation of the Act was vested with officials supportive of impact assessment and the regulatory regime put in place by the Act was insulated from the government's fiscal restraint policies.
- *Functions*: under the new Act, the perspective on BC Hydro's regulation changed from one of passive endorsement to a more active enforcement of impact assessment and rate reviews.
- *Administrative structures*: the Act established the BC Utilities Commission, transferred significant responsibilities for impact assessment to two existing government ministries and eventually led to a significant internal reorganization within BC Hydro.
- *Processes and mechanisms*: the Site C review was the first under the new Act and as a catalyst for change it established the manner by which the new provisions for impact assessment review were to be implemented in practice. This included opportunities for extensive public participation and a broadening of the terms of reference for project reviews.
- *Culture and attitudes*: both the Act and the Site C review reflected a wider sense of public unease regarding the province's energy planning. BC Hydro was seen to be unaccountable and out of control. The Site C review brought the utility back into line and showed the potential for change inherent in the enabling legislation. However, once the initial catharsis had occurred, political attitudes retrenched somewhat and the view emerged that the reforms, perhaps, had gone too far.

The response to this last point is to suggest that direct political decision making might be necessary to circumvent some procedures. This would constitute a further adjustment to the legitimation leverage point and would, in turn, initiate a further series of adjustments throughout the remaining leverage points.

Impact asessment as the target of reforms

The perceived failure of the BEARP was one element contributing to calls for the abandonment of the Canadian government's review process for environmental assessment (EARP) in the early 1980s. The latter had been initiated in 1974 as an administrative procedure. It lacked legislative backing and was reliant upon proponent compliance for implementation. By the end of the 1970s, the process was being ignored with impunity by many agencies and needed revision (Fenge and Smith, 1986; Rees, 1980). Initial proposals called for radical changes that would have led to the replacement of EARP with a newly legislated federal process for impact assessment with wider application and clearer guidelines for implementation.

However, for opponents of reform, the application of EARP to developments in the North was seen to be redundant on the basis of duplicity with industry assessments and the actions of the National Energy Board. For example, EARP was seen by the hydrocarbon industry to be unstructured, time-consuming, repetitive and unnecessary. This view was endorsed by several government committees and a variety of federal departments 'reluctant to increase EARP's authority at their own expense' (Fenge and Smith, 1986: 599). The BEARP failings added considerable credence to this view and were instrumental in the derailment of any large-scale reforms to EARP. Instead, a reform package was approved in 1984 that modified the application of the original process and added some formal guidelines for implementation (Fenge and Smith, 1986). These modifications ultimately proved to be insufficient and, by the onset of the 1990s, reform of EARP was once more on the political agenda (Smith, 1990b).

Impact assessment as a function of international aid

Generally, developing countries face several constraints regarding suitable institutional arrangements for impact assessment (OECD, 1989; Kennedy, 1988b). These include:

- insufficient political awareness of impact assessment
- insufficient public participation
- poor or absent legislative frameworks
- absence of an institutional basis
- insufficient skilled human resources
- inadequate baseline data
- inadequate financial resources.

Consequently, most developing countries have relied upon international agencies to help initiate impact assessment provisions.

Horberry (1985) has indicated that the assistance of international agencies regarding impact assessment reflects a desire to augment the efforts of national environmental policy makers and administrators through:

- *Political support*: international agencies lend legitimacy to environmental policies and elevate the political status of administrators.
- *Institutional strengthening*: funds, expertise and training opportunities are used to strengthen the institutional capacity of an agency to implement impact assessment.
- *Technical capability*: assistance is extended in a variety of forms ranging from consulting assistance to guidelines and training courses.

However, his analysis indicated that such aid is often shaped as much by the characteristics of the donor agency as it is by the needs and priorities of the recipients (Horberry, 1985). Moreover, Kennedy (1988b) concluded that impact assessments funded by aid agencies were not comprehensive in their content and generally did not satisfy domestic requirements.

The developments in both Amazonia and Bangladesh highlight the role of international donor agencies (especially the World Bank) in the implementation of impact assessment in developing countries. These examples indicate that the role of agencies such as the World Bank is crucial, particularly if the political will to protect the environment, aboriginal peoples and/or social values is absent. Developments in the Amazon are testimony to the weakness of the Bank's previous commitment to impact assessment. The proposals in Bangladesh and their eventual outcome will reflect on the World Bank's present policy on the environment and its vision of sustainability.

Summary

In all three scenarios, the pivotal leverage point appears to be that of legitimation. Where impact assessment for frontier developments is lacking or weak, it reflects a more generalized absence of firm environmental legislation and administrative agencies with strong environmental mandates. Both are symptomatic of political diffidence. Consequently, there is an unwillingness to commit finances to 'review procedures' that are perceived to be unnecessary, time-consuming, duplicative or just plain obstructionist. Such is the reputation of impact assessment as it has evolved over the past 20 years.

In response, numerous efforts have been made worldwide to introduce new provisions for impact assessment and to update existing ones. These range from the adoption of a directive on impact assessment by the EC; to the initiation of impact assessment procedures in countries such as Indonesia, the Philippines, India and Thailand; and, to the revision of institutional arrangements for impact assessment in New Zealand and Canada. However, for this situation to continue to change, impact assessment must become more than an intrusive and unwelcome review mechanism. Institutional arrangements are required that provide for impact assessment as an approach to planning, as a means not so much of project approval but of project design. Only then will impact assessment for frontier developments become a reality.

Interest representation

Frontier developments invariably involve decision making at a normative planning level. As with institutional arrangements, the issue of legitimation is again paramount. The basic barrier appears to be that of being heard. Decisions on frontier developments are perceived to be a political privilege. For this reason, lobbying tends to be the dominant form of influence upon decisions affecting frontier developments. The pressure exerted by environmental groups in the United States which resulted in policy changes by the World Bank illustrates the effectiveness that lobbying can have on decision making. However, lobbying is only effective in countries where there is a strong pluralistic tradition in politics. Thus, an environmental lobby related to development of Amazonia is inherently more effective in the United States than similar efforts in Brazil itself.

Aside from the practice of lobbying, formal approaches to interest representation at a normative level are rare. As a result, *ad hoc* approaches tend to be used. Often, a proposed development is so politically controversial that a major examination is initiated 'to mobilise public support for a decision and the policies that envelop it' (Kemp, O'Riordan and Purdue, 1984: 477). This was the case in Canada with the Berger Inquiry (Smith, 1987a), in Britain with the Sizewell B Inquiry (O'Riordan, Kemp and Purdue, 1988) and in Australia with the Ranger Inquiry (Formby, 1981).

Regardless of the approach used, interest representation on frontier developments is characterized by large differences in the nature of the concerned stakeholders and a fundamental divergence in the various goals and objectives for their participation. Frontier developments are large scale and capital-intensive. They originate with a desire for economic growth and/or social benefit from the perspective of the dominant society that usually has initiated the proposed development. However, frontier developments frequently impact upon remote societies, highlighting the problems of disadvantaged sectors and the treatment of aboriginal peoples. The goals and objectives of these stakeholders often show vast differences in values from those of the dominantly urban population that benefits most from the development. Berger's (1977) characterization of a 'resource frontier' or 'aboriginal homeland' succinctly encapsulated these contrasting viewpoints.

There is no one way for these issues to be tackled. But to ignore their existence or to continue to rely upon *ad hoc* mechanisms for response, is to perpetuate the oppression of some stakeholders to the advantage of others. Impact assessment can only mature as a democratic approach to environmental planning by giving explicit consideration to the need for interest representation in decisions on frontier developments.

Impact assessment

The four case studies highlight the intrinsic difficulties encountered by impact assessment at a normative planning level. For example, the BEARP showed that the absence of a specific project can result in an assessment that lacks focus, with little influence upon policy implementation. In contrast, the Site C evaluation illustrated how a specific project appraisal at a normative level can function as a *de facto* review of policy on frontier development. In neither instance was the scoping adequate to specify the impacts to be assessed, the boundaries for assessment nor the range of alternatives to be considered.

Uncertainty and the problems of impact prediction have complicated efforts to resolve the flooding problems of Bangladesh. How reliable is the science of prediction in situations where the absence of knowledge is matched only by the scale of the problems incurred and the paucity of resources devoted to their resolution? The dilemma here is that while uncertainty and data limitations are increased by project scale, so is the need for an accurate prediction of impacts. Without good predictive measures, the assessment of significance loses credibility and frontier developments are approved with little sense of their potential effects. This has certainly been the case in Amazonia, where the size of the development and the area it affects have been used as an excuse for insufficient planning and inadequate impact assessment.

These problems are not unique to the four case studies. An international seminar reviewing aid agency experience with impact assessment (OECD, 1989), found that scoping and a paucity of information to guide impact prediction and significance assessment were critical difficulties characteristic of frontier developments. However, while these problems may be viewed as endemic, it is possible also to view them as symptomatic of the fact that most methods for impact assessment have been developed for application at a project-specific level that is not always compatible with the wider policy ramifications and impacts of frontier developments. The experience with the BEARP illustrates this point well, and highlights the need for impact assessment methods tailored to the demands of frontier developments. Adaptive methods are needed that:

- provide for scoping, prediction and significance assessment
- can be applied in situations where data are sparse and unreliable
- focus upon wider policy implications and ramifications.

Conclusion

At the onset of the chapter it was suggested that frontier developments pose problems for impact assessment related to:

- project justification

133

Frontier developments

- terms of reference
- uncertainty
- impact assessment science.

The review of the case studies through the application of the integrative framework for sustainable resource management suggests that the potential for redefinition of impact assessment of frontier developments rests with:

- recognition that the absence of political commitment and adequate institutional arrangements are major barriers to the application of any impact assessment procedures to frontier developments
- formal and explicit recognition of the need for stakeholders, especially impacted groups, to have representation within the planning and decision-making processes for frontier developments
- the implementation of appropriate methods for scoping, prediction and significance assessment of frontier developments.

CHAPTER 8

Linear facilities

Introduction

Resource trade is an important basis for contemporary economic structure in the world. Linear facilities are the physical manifestation of this reality, providing for the transportation of resources from the source of supply to centres of demand (e.g. transmission lines and pipelines) and the very basis for transportation itself (e.g. highways and railways). Linear facilities exist as networks, usually on a regional scale and tend to be a highly visible, tangible and intrusive component of the landscape.

Linear facilities are the infrastructural prerequisites to industrialized society and they are often perceived to be synonymous with progress and development. However, linear facilities tend to shape society around them, altering landscapes and patterns of development. As such, linear facilities typify a technocratic society and they frequently are a good illustration of society and landscapes conforming to technology rather than technology operating within the confines determined by society.

Given their nature, linear facilities have been a major concern within impact assessment. This chapter commences by briefly outlining the concerns and focus for impact assessment in the siting of linear facilities. The integrated framework for analysis is then applied to the case study of transmission line planning in Ontario to illustrate the relative strengths and weaknesses of existing impact assessment practice regarding linear facilities. The chapter concludes with a discussion of changes in impact assessment to improve the ability to predict, assess the significance of and evaluate the impacts of linear facilities.

Siting linear facilities

Weary (1984) suggested that because of its inherent complexity, linear route selection requires a detailed and formal approach to planning. This would include regional screening, the identification of alternatives, the evaluation of those alternatives, and final site selection. However, despite 'the obvious advantages of this type of methodology . . . the majority of . . . routing studies adopt a less rigorous approach' (Weary, 1984: 8). Increasingly, this lack of rigour has led to delays in project approval and even to project cancellations. Based on the US experience, Massa (1984) indicated that such project failures result from routing studies that:

- have a weak definition of project need, objectives and/or characteristics
- reveal an absence of, or inadequate, interest representation
- fail to follow a clearly traceable siting process.

She noted that the 'core of the siting process involves the successive narrowing of an appropriately identified region' (Massa, 1984: 17) and proposed a phased process for logical and comprehensive route selection involving:

- project goal formulation
- definition of technical characteristics and requirements for the project
- identification of required and potential relationships between the project and the environment
- identification and screening of candidate areas
- evaluation of candidate sites.

Thus: 'As candidate areas and candidate sites are selected and evaluated, the analysis becomes progressively more detailed, using selection criteria derived from the previously established project objectives and environmental analysis' (Massa, 1984: 17).

One widely advocated approach to linear facility siting is the use of corridors for the routing of transportation and resource delivery systems. Weir and Reay (1984: 122) defined the corridor concept as: 'the placing of transportation facilities within a planned continuous strip of land . . . in such a manner that the use of land for right of way within the corridor is maximized, while . . . mitigating against any adverse environmental or social effects on adjacent land uses'. Corridors are therefore intended to reduce land-use conflicts by removing the need for multiple route alignments. In effect, many linear facilities occupy adjacent rights of way forming a *de facto* utility corridor. This has several advantages including:

- conservation of land
- facility integrity
- limited environmental disturbance
- land-use planning co-ordination.

However, utility corridors can have several, potentially large, disadvantages arising from:

- increased disaster potential
- the concentration of impacts and the creation of cumulative effects

- increased construction costs owing to the proximity of other facilities
- land underutilization.

Typically, utility corridors link some combination of the transportation system (a railway and/or a highway), with elements of the resource delivery system, such as a pipeline or transmission line. Each component of a corridor has its own restrictions affecting location and its effects on the environment can be quite distinct:

- *Railways*: the least flexible component in terms of locational requirements. When included in a corridor, a railway will dictate the whole corridor alignment. Their impact is similar to that of a highway but of a lesser magnitude because of their reduced right-of-way width.
- *Highways*: cause total environmental disturbance locally and have the greatest impact of any component both during construction and facility operation. Long-term impacts can be reduced by reclamation and landscaping.
- *Pipelines*: environmental disturbance during construction is high but mitigation practices can minimize long-term impacts. Engineering constraints are few and relate to safety, reliability, economics, construction priorities, aesthetics and environmental factors.
- *Transmission lines*: create the least physical disturbance during construction, have the fewest engineering constraints and permit the most effective reclamation practices. Their major impacts are visual, electrical interference and perceived health risks.

In addition to being standard elements within a utility corridor, these linear facilities have been considered on an individual basis within impact assessments. However, because most rail infrastructure in the world was put in place prior to 1950, proportionally more focus has been placed on routing selection and its implications regarding highways, pipelines and transmission lines.

Highways

Highway location is subject to fairly rigid parameters: road hierarchy, population density, projected volumes of traffic, expected speed of traffic, safety and sight distances (Weir and Reay, 1984). Consequently, highway alignments are dictated by:

- the location of traffic generation centres
- gradient requirements
- the need for a stable, well-drained substructure
- construction material proximity and availability
- aesthetic concerns.

The immediate impacts of highway construction on the biophysical

137

environment include visual impacts on the landscape, loss of habitat and habitat alteration, such as hydrological changes and species composition changes. But a major road can also have severe long-term effects. Based upon their studies in Britain, Sheate and Taylor (1990) have indicated that some of the long-term implications of major road developments on adjacent biotic communities are:

- *Pollution*: including increased runoff, soil erosion, sedimentation, noise, emissions and salt deposition.
- *Biogeographical isolation*: highways are effective isolating mechanisms. The physical separation of biotic communities into smaller areal units can reduce species diversity and threaten ecological integrity
- *Aerodynamics and hydrology*: including increased risk of wind-blow effects, exposure of communities and effects on groundwater levels and flows.

This combination of rigid locational requirements and potential ecological havoc has made highways a frequent subject for impact assessment. Indeed, when impact assessment was still in its infancy, McHarg (1969: 31) had summarized the problems posed by highway construction as follows:

In highway design, the problem is reduced to the simplest and most commonplace terms: traffic, volume, design speed, capacity, pavements, structures, horizontal and vertical alignment. These considerations are married to a thoroughly spurious cost–benefit formula and the consequences of this institutionalized myopia are seen in the scars upon the land and in the cities.

He also suggested that 'a major highway presents an excellent opportunity to demonstrate that natural processes can be construed as values in such a way as to permit a rational response to a social value system' (McHarg, 1969: 31). For McHarg this meant that the best route was that with the highest overall social benefit, and the only way that could be determined was by mapping valued environmental components as constraints on location. This was the premise McHarg pursued in his study for the Richmond Parkway in New York City to demonstrate the utility of his overlay approach to impact assessment.

However, in many countries, highway construction programmes were well entrenched prior to the advent of impact assessment and it was only after much controversy that the situation was changed. Perhaps the best illustration of this point is Britain where a series of disputes in the mid-1970s led to the disruption of many public inquiries and a review of the basis upon which motorway (expressway) need was predicted (Tyme, 1978). The disputes involved new highway construction and the attempts by intervenors to have the question of need open to scrutiny rather than just the route of the proposed road.

Britain's first motorway had opened in 1958. It was the manifestation of a master plan developed in 1946 whereby motorways were seen as establishing the framework upon which the nation's future town and country planning could be based by supplying a pattern of principal national routes (Starkie, 1982). Opponents of Britain's motorway programme criticized highway construction on philosophical and political policy grounds, contending that highways generate congestion, reduce transportation options, destroy housing, socially impact on communities, monopolize scarce resources and commit communities to 'dispersal planning' (Tyme, 1978).

The protests led to a review of road planning in Britain. It revealed that traffic forecasts were inaccurate and speculative, and confirmed that the reliance upon benefit–cost analysis had precluded consideration of environmental factors (Starkie, 1982). The review led to significant changes in the motorway programme which resulted in many projects being downsized, with a shift away from motorway construction towards more emphasis upon bypasses to avoid urban areas. The controversy over highway construction was also a major impetus in the development of a formal manual for impact assessment in Britain, and consistent with EEC Directive 85/337, impact assessments are now required for major highways in Britain under regulations within the Highways Act.

The EEC Directive has also influenced the development of impact assessment for highways elsewhere in Europe. For example, the Transjurane N16 in Switzerland generated a number of concerns, including the impact of increased air emissions at an important junction in the town of Bassecourt. The road was used to help develop and refine an impact assessment methodology that could be used for subsequent projects in Switzerland (Hertig, 1990). In Portugal, the repercussions of the EEC Directive led to that country's first ever impact assessment, for a highway near Lisbon (Canales, 1989). The main concerns were the road alignment and its effects upon landscape quality in an important tourism area. The project was highly controversial and had been postponed several times. The impact assessment helped clarify the effects that the road would have and showed that while some resettlement was necessary, the major impact of increased noise levels could be mitigated through some design modifications. Moreover, the road would significantly reduce traffic levels on existing routes, both decreasing pollution levels and adding to the area's attractiveness to tourists (Canales, 1989).

Thus, in both these cases the experience with impact assessment was considered positive and showed signs of improvement over the situation discovered by Kennedy (1988a) in his review of the experience with impact assessment in North America and Western Europe. The review was based on a cross-national study by a special United Nations task force utilizing 11 case studies of impact assessment for highway projects and dams in 6 different countries. The findings indicated that impact assessment should become a more integral component of project planning, that the scientific quality of most impact statements was poor, and that impact prediction was particularly weak and reliant upon unsubstantiated professional judgements, rather than any verifiable methodology (Kennedy, 1988a). Thus, it appears that while highways have been a frequent first target for impact assessment initiatives, the methodology for their assessment is by no means an accepted, refined standard.

Pipelines

Pipelines have a large construction impact. They are sensitive to soil and topographic features, with the ideal routing being the shortest distance between source and terminal point. Actual alignments are influenced by the product being transported, pipeline pressure, topographical and soil conditions, safety concerns and required surveillance and control procedures (Weir and Reay, 1984). Gradients are not a major obstacle, but the timing of construction, river crossings and environmentally sensitive areas are major considerations regardless of the size of pipe involved.

Jakimchuk (1987) gave a good illustration of this last point when he reviewed the impacts of the Norman Wells oil pipeline in Canada in comparison with the effects of the Alyeska oil pipeline in Alaska. The two projects were quite distinct from one another in their characteristics, the Norman Wells project involving a buried, 324 mm oil pipeline built between 1984 and 1985 for a distance of 866 km from Norman Wells to Alberta, while the Alyeska pipeline is a partially buried, partially elevated large-diameter oil pipeline (762 mm) built between 1974 and 1977 for a distance over 1200 km from Barrow to Valdez in Alaska. Despite their physical differences, the two pipelines instigated similar environmental impacts including siltation, erosion, slope instability and effects on wildlife, fisheries and their habitat. In addition, both had a common problem with permafrost affecting pipeline integrity and terrain sensitivity. The corollary here is that pipeline impact is a function of terrain disturbance and the linear character of the facility. Such factors cannot be mitigated by merely reducing the dimensions of the project.

Impact assessments for pipelines are a routine component of most siting studies, but it is unclear how effective these assessments have been either in project approval or in actual implementation of mitigative measures. For instance, Zallen et al (1987) analysed the assessment of three linear developments in the Coquihalla Valley of British Columbia:

- a proposed 762 mm oil pipeline
- a 914 mm natural gas pipeline
- a four-lane highway.

They found a wide variation in the impact assessment characteristics for the three projects. Neither pipeline project impact assessment indicated the type nor the magnitude of predicted impacts very well, and all of the assessments described impacts with imprecise terms such as 'minimal', 'high' or 'severe'.

Moreover, cumulative impacts were not addressed. The major impacts were found to relate to stream crossings (e.g. encroachment, diversions, sedimentation, effects on fish spawning and water quality degradation) and the effects of construction activity on habitat (e.g. scheduling of construction activities, sedimentation, erosion and habitat encroachment). Zallen et al (1987) concluded that mitigation was possible but that it required site-specific plans and monitoring to be effective. In effect, the study showed that the impact assessments were placing a reliance upon mitigation and monitoring in lieu of conducting effective prediction, significance assessment and evaluation within the actual assessments themselves.

Evidence in support of this conclusion can also be drawn from a study conducted by Moncrieff et al (1987). They examined a series of 15 pipeline projects in southern Ontario from 1974 to 1985 to determine if the treatment of environmental issues had improved during that 10-year period. Performance was assessed with respect to:

- route selection
- environmental evaluation and impact prediction
- mitigation procedures
- monitoring
- costs.

Moncrieff et al (1987) found that information on route selection was generalized. Specific details on the process used were absent and routing relied upon data collected at a scale (1 : 50 000) that was too small to provide sufficient detail to evaluate alignment impacts. However, improvements were found once routing was subject to increasing scrutiny as a consequence of rising levels of interest representation in the regulatory review of proposals. The environmental assessments showed increasing amounts of detail in accordance with provincial guidelines on pipeline planning and site-specific evaluations were seen to have benefited from increased landowner involvement. Major issues involved habitat loss, land-use trade-offs and construction activities, but many assessments still relied upon *professional judgement* as their criterion for determining impact predictions and significance. In most cases, this judgement was unsubstantiated and amounted to little more than an informed guesstimate of likely effects.

Mitigation was found to be problematic in most assessments. The key aspects related to rarely documented issues, such as the environmental attitude of the proponent and its contractors, their commitment to planning and their communications strategy. The most effective form of mitigation was considered to be avoidance through appropriate route selection and timing of construction (Moncrieff et al, 1987: 153), but most monitoring reports provided little information on the effectiveness of mitigation measures. Lastly, it was found that many mitigative measures were related to the scheduling of activities and involved no direct costs. Others reflected good construction practices and were insignificant in their cost, while others actually realized cost savings through reduced expropriation payments. The implication here is that cost should not be held as a significant constraint to effective mitigation as successful mitigation can also benefit the proponent's own bottom line.

Transmission lines

There are few absolute constraints on transmission line siting from a technical perspective (Weir and Reay, 1984). However, location is dictated by the whereabouts of load centres and generation sites, which effectively constrain the routing options. The cheapest route is usually a straight line but this can be modified by the desire to minimize land-use conflicts through the use of existing rights of way and the avoidance of sensitive areas. Rolling or undulating terrain can also be used as a visual screen to reduce the aesthetic impact on vistas. The location of substations to step down power from transmission lines is more restrictive and is a function of load centre location (Weir and Reay, 1984).

Because rerouting options can often be quite limited, electric utilities have been at the forefront of efforts to improve compensation mechanisms as mitigative measures for transmission lines. For example, Thompson (1984) documented two case studies in Montana where compensation was used as a mitigative measure rather than the rerouting of a proposed transmission line. This study suggests that compensation for powerline impacts can be a cost-effective alternative to relocation.

However, many transmission line impacts (either real or perceived) may not be suited to monetary compensation. Some factors are intangible, while others are just difficult to quantify. Concannon (1984) illustrated this point well when she outlined the public involvement strategy used by the Bonneville Power Administration (BPA) in locating transmission lines in the US Pacific North-west. The BPA experience indicated that transmission lines generate the following list of public concerns:

- visual and aesthetic impacts
- health risks
- project need and benefits
- energy conservation and its ability to postpone or cancel the project
- impacts on heritage sites
- effects on natural resources
- precedents for future corridor location
- weighting of impacts
- economic impacts
- land-use conflicts
- safety
- public involvement in decision making.

142

Many of these concerns are subject to speculation as the scientific evidence itself is either absent, speculative or contradictory, as in the case of the perceived health risks associated with high-voltage transmission lines. In these circumstances, the affected public have no firm basis upon which to fix a level of compensation, and impact prediction gives way to risk estimation as risk assessment supersedes impact analysis.

Given this level of uncertainty and the relatively fixed nature of transmission line location near to population centres, it is not surprising that transmission line routing has been the subject of frequent public controversy. In response, electric utilities have had to alter their approach to transmission line planning. Some have embarked on this course willingly, others with a conspicuous lack of enthusiasm.

Ducsik (1984) suggested that utilities use a siting process incorporating four hierarchical activities, usually in sequence:

- *reconnaissance and screening*: a specified methodology and criteria applied to a range of siting options on a regional basis to focus the evaluation to a limited number of alternatives
- *comparative evaluation*: the identification of a preferred alternative through a systematic comparison of the options
- *detailed assessment*: site-specific planning for the preferred option
- *review and implementation*: regulatory scrutiny and approvals.

He then examined the rate of progress among utilities regarding the adoption of public participation in siting planning, citing the benefits of a participatory approach in comparison to the traditional and paternalistic DAD (decide, announce, defend) approach used by utilities until the mid-1970s. His analysis suggested that utilities were slow in responding to the challenges of public participation and that while a few utilities were progressive in their planning, the vast majority were not (Ducsik, 1984).

One utility identified by Ducsik's survey as progressive was Ontario Hydro. This finding corroborates the work of Smith (1983, 1984b), who examined public participation in electric power planning in Ontario. His studies documented Ontario Hydro's responsiveness in the area of public participation and illustrated how that was reflected in a revised approach to transmission line routing. However, Smith (1983) also showed that widespread public participation in power planning is an illusory goal until the siting phase is reached. It is only at the site-specific level that strategic issues come into focus and become understandable, tangible and real to lay audiences. Hence, the controversy and high level of protest that often accompany public involvement efforts in siting studies arise when the public is asked the question 'do you want *x* facility in your front yard or your backyard?' to which a common response is a questioning of the need for the project in the first place, a question utility planners answered far earlier during the system planning stage (at which point little public interest existed in the planning effort).

Public participation in utility planning continued to evolve throughout the 1980s (e.g. Smith 1988a), and by the 1990s progressive utilities were using a variety of approaches to elicit interest representation in their system planning efforts (e.g. Ontario Hydro's 1990 supply/demand study). However, two factors continue to constrain these efforts: (1) lay audiences still tend to respond better to site-specific issues despite the efforts of utilities to improve their planning efforts; and, (2) the level of uncertainty regarding some perceived impacts of transmission lines (such as long-term health risks) has continued to rise even as the level of research on the topic has increased.

Summary

These examples show that it is the strategic level of planning that provides the context for impact assessment of linear facilities. It is at this level of planning that alternatives are evaluated and choices are made concerning the actual nature of the facility to be built. Strategic planning is also the level where impact assessment is most readily understood and accepted as it emphasizes the prediction of impacts, the determination of their significance and an ultimate evaluation of their effect. All these tasks are central to the development of impact assessment methods and methodology (Thompson, 1990).

For linear facilities, the major decision involves route selection and the potential for impact mitigation through good strategic planning. Therefore, the key questions to be addressed in the impact assessment of linear facilities are:

- *Problem identification*:
  · Has project need been addressed?
  · What are the objectives of the project and its essential characteristics?
- *Scoping*:
  · Have valued environmental components been identified?
  · How have spatial and temporal variations in components been taken into account?
  · What alternative routes are being considered?
- *Impact prediction*:
  · What methods were used to predict possible impacts?
  · Have both construction and operating impacts been addressed?
  · How has the issue of uncertainty been addressed?
- *Impact significance*:
  · What criteria have been employed to determine significance?
  · Have the cumulative effects of a utility corridor been assessed?
- *Impact evaluation*:
  · How have the various factors been weighted in making trade-offs between differing land-uses?
  · How many alternatives have been evaluated?

· What provisions have been made to monitor impacts?
● *Impact mitigation*:
· Have the project's impacts been successfully mitigated through effective routing?
· What compensation package is required?
● *Interest representation*:
· How open should the siting process be?
· What provisions are there for stakeholder input into the determination of impacts, their significance and their ultimate evaluation?

The relative strengths and weaknesses of existing approaches to impact assessment for linear facilities can be determined by reviewing these questions relative to the experience with locating bulk transmission facilities in southern Ontario. The example benefits from, and reflects changes arising out of, the background in Ontario with transmission line planning during the period from the late 1970s to the late 1980s. This topic has been the focus of an ongoing research programme, with previous findings published by Smith and Cattrysse (1987) and Smith (1983, 1984a). In addition, the present case study draws heavily upon information contained in Ontario Hydro's 1990 environmental assessment for bulk transmission facilities west of London, Ontario.

Case study: transmission line planning in southern Ontario

Problem identification

At the end of the 1980s, transmission facilities in the London–Sarnia–Windsor area of southern Ontario (Fig. 8.1) were inadequate to ensure a secure and reliable power supply. In addition, the 1988 load forecast had predicted an annual increase in demand of 1.7 per cent through to the year 2015, further emphasizing the need for improved bulk transmission facilities. In response to this demand, Ontario Hydro proposed the construction and operation of a 500 kV transmission line from London to Sarnia and a 230 kV transmission line from London to Windsor (Ontario Hydro, 1990). The routes recommended by the utility would mean a 76 km line from London to Sarnia and a 148 km line between London and Windsor.

The purpose of Ontario Hydro's proposal was to meet the electricity requirements of the London–Sarnia–Windsor area to the year 2015, while maintaining interconnection capabilities between Ontario Hydro and the Michigan Electric System and improving the overall reliability and stability of the Bulk Electricity System (BES), the province's electric power grid. Specifically, the proposed undertaking involved the planning and design of overhead alternating current transmission lines, acquiring the necessary property rights for them, and constructing, operating and maintaining them and related support facilities (Ontario Hydro, 1990: 3). The required facilities included:

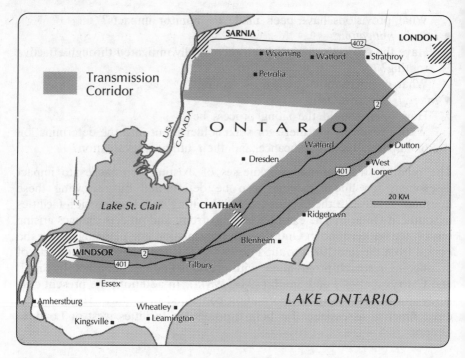

Fig. 8.1 Transmission corridors in south-western Ontario

- one 500 kV single transmission circuit from the Longwood transformer station (TS) near London to the Lambton generating station (GS) south of Sarnia
- two 230 kV transmission circuits between Cowal Junction south of the Longwood TS and the Chatham switching station (SS)
- two 230 kV transmission circuits between the Chatham SS and the Windsor Lauzon TS
- associated transformation, switching, communications, control and local transmission facilities necessary to integrate the undertaking into the overall Ontario Hydro BES.

Institutional arrangements

Ontario Hydro's proposal was submitted to the Ontario Ministry for the Environment (MOE) for review and approval under the Environmental Assessment Act (RSO 1980: c. 140). The Act requires proponents of development proposals, or 'undertakings', to prepare and submit an 'environmental assessment' for formal government approval. As such, the Act provides the legislative basis for impact assessment in the province.

The stated purpose of the Environmental Assessment Act (s. 2) is for the 'betterment of the people of the whole or any part of Ontario by providing

146

for the protection, conservation and wise management in Ontario of the environment'. To achieve this objective, the Act establishes the basis for environmental assessment, providing for the involvement of the public and government ministries in the development planning and approvals process. Unless exempted, the Act applies to all provincial ministries and agencies, including Ontario Hydro.

Within the Act, an undertaking is defined as an enterprise or activity, or a proposal, plan or programme in respect of an enterprise or activity, by or on behalf of the government of Ontario, a public agency or a municipality. Environment is also broadly defined by the legislation as including physical features, biological subjects and human and ecological systems, and any 'part or combination of the foregoing and the interrelationships between any two or more of them' (s. 1(c)).

The Act requires proponents to ascertain whether or not an undertaking is subject to environmental assessment. Where it is, the proponent is required to use the following procedure:

- An environmental assessment (EA) must be prepared by the proponent and submitted to the Minister of the Environment. The EA should be prepared after consultation with the MOE. Proponents are encouraged to involve the public in the preparation of the EA.
- The submitted EA is then circulated by the MOE for review by government agencies. A co-ordinated review is prepared and released along with the EA for a minimum of 30 days for public comment.
- The EA may then be (1) accepted by the Minister and approved, or (2) accepted by the Minister but be the subject of a formal public hearing by the Environmental Assessment Board, or (3) subject to a formal hearing by the Environmental Assessment Board to determine its acceptance and approval. In each instance, the provincial Cabinet has ultimate authority to approve or alter the final decision regarding a project.

The Act contains specific requirements and procedures for public hearings. It requires that hearings must be held by the Environmental Assessment Board in response to any written request for a hearing received by the Minister while an EA is open to public review.

Under the Act, the required components of an environmental assessment are:

- a description of the purpose of the undertaking
- a description of and a statement of the rationale for:
  · the undertaking
  · alternative methods of carrying out the undertaking
  · alternatives to the undertaking
- a description of:
  · the environment affected, directly or indirectly

· the effects on the environment of the undertaking
· actions necessary to change, mitigate or remedy those effects
● an evaluation of the environmental advantages and disadvantages of the undertaking and its alternatives.

Proponents are prohibited by the legislation from proceeding with an undertaking until formal approval has been granted. Through all these procedures, the Act seeks to promote comprehensive study and planning, informed decision making and review, and protection of the environment.

Ontario Hydro is subject to the provisions of the Environmental Assessment Act as it is a provincially owned utility. Its responsibilities include the supply of electricity in Ontario in a manner reflecting the social, economic and environmental aspirations of the province. In addition, Ontario Hydro's traditional mandate under the Power Corporation Act (RSO 1970: c. 354) has been to provide 'power at cost'. Consistent with this philosophy, Ontario Hydro generates the majority of its electricity using nuclear power, with a significant proportion of its base load supplied by hydro generation and a somewhat smaller contribution from thermal generating stations. The utility is one of North America's largest and it has a well-deserved reputation for being an innovative and responsive power utility.

Interest representation

Ontario Hydro's environmental assessment for the bulk transmission facilities west of London included pre-submission consultations with 19 different provincial government ministries. In addition, the utility held five pre-submission consultation meetings in Toronto for government reviewers between December 1987 and April 1990. These efforts were supplemented by an extensive public involvement programme that included contributions from municipalities, special interest groups and the lay public.

A variety of approaches was used to elicit public input into the EA, including:

● project newsletters
● a media programme
● information centres
● public meetings as requested by specific groups or property owners
● presentations to local councils and interest groups
● review groups
● a 24-hour telephone information line.

Of these approaches, the review groups are particularly noteworthy. Continuing the practice it had initiated with earlier routing studies (Smith and Cattrysse, 1987), Ontario Hydro established public review groups to provide opportunities for more in-depth commentary on, and contribution to, the siting process. Two groups were established. One involved planning representatives from differing local, regional and provincial agencies. The other involved representatives from various regional, local and provincial interest groups. Among the groups specifically involved through this approach were a number of agricultural federations, associations for various agricultural products, a range of naturalist and environmental organizations, various business groups and the local chambers of commerce.

Both review groups met concurrently throughout the planning process. Among the topics addressed by the groups were:

- the need for new transmission facilities
- factors to be used in identifying route corridors
- environmental planning objectives
- the environmental ranking process
- constraint ranking
- corridor identification
- feedback from information centres
- corridor adjustments
- development of alternative routes
- determination of preferred routes.

In this manner, the review groups acted as a 'sounding board' for Ontario Hydro's planners and they were an integral component of the impact assessment methodology employed by the utility.

Impact assessment

Ontario Hydro's route selection planning process

The first step in Ontario Hydro's planning process involved the selection of a preferred alternative to satisfy the purpose of the undertaking. Following the requirements in Ontario, this meant an initial screening of alternatives to determine project need. Thus, an initial range of options was considered, including:

- the null alternative (do nothing)
- conventional generation options
- supplemental generation options
- purchase alternatives
- transmission alternatives.

These options were evaluated on the basis of their cost, timeliness, technical merit and net environmental impact. Based on this initial screening, the option to add new transmission facilities was selected as the preferred alternative for further study. Computer simulation modelling was then employed to determine which of three possible combinations of new

149

transmission facilities was preferred using the same criteria as for the initial screening. This analysis indicated that the proposed combination of a 500 kV transmission line from Longwood to Lambton and a double circuit 230 kV line to the Windsor area was preferable as it had less net environmental impact and a lower capital cost than the others. At this juncture, the impact assessment became focused on the location of a preferred route for the siting of the proposed transmission facilities.

Ontario Hydro's route selection study process involved the integration of environmental, socio-economic and engineering components with the contributions from the comprehensive public involvement programme. The first stage of the study process involved the determination of 'route corridors' to reduce the study area (Fig. 8.2). A corridor is defined as a 'continuous linear region generally located through areas of relatively low

Alternative corridors

PHASE 1

- Collect data within the study area at scale 1:50,000 and computer store using 250 m cells
- Determine and map environmental factors
- Determine priorities and prepare constraint maps
- Identify alternative transmission corridors

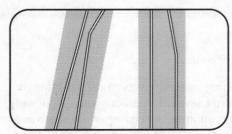

Alternative routes within identified corridors

PHASE 2

- Collect data within alternative corridors and zones at scale 1:10,000
- Prepare composite maps
- Identify alternative routes
- Evaluate and compare alternatives

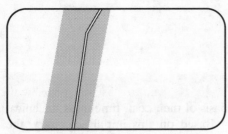

Recommended corridor, recommended route

PHASE 3

- Identify preferred routes
- Select recommended routes

Fig. 8.2 The route selection process

constraint, where transmission routes will have the least environmental, social and technical impact' (Ontario Hydro, 1990: 79). Thus, corridors were distinguished by identifying areas of high constraint while retaining areas of lower constraint. From the perspective of impact assessment methods and methodology, the steps involved in this process were:

- scoping
  - · study area identification
  - · data collection at a scale of 1 : 50 000
  - · identifying environmental situations
- prediction
  - · determining environmental planning objectives
- significance assessment
  - · ranking of objectives
  - · environmental constraint maps
  - · identification of corridors

The second phase of the route siting study was the identification of alternative routes. This phase can be viewed as a second iteration of the steps used to identify the route corridors, but at a more refined scale. Thus, the identification of alternative routes involved data collection at a scale of 1 : 10 000 and the use of environmental priorities to produce a composite map to illustrate site-specific constraint information. Finally, the third phase of the route study process was that of evaluation. In this phase, the alternative routes were evaluated to select and identify the preferred route (Fig. 8.2).

Scoping

The initial task was the delineation of the study area. Using the terminal points of the required facilities, the study area was divided into two: a northern portion linking the Longwood TS to the Lambton GS, and a southern portion from the Longwood TS to Windsor (Fig. 8.1). The objective at the corridor identification phase was to eliminate potentially high impact areas and to focus the study on those areas with a lower probability for impact.

To assist in the determination of these constraint areas, data were collected at a scale of 1 : 50 000. This scale was employed as it is commonly used for resource mapping and planning by other government agencies and these sources could then be used in the compilation of the necessary data. At this scale, mapping is generally confined to areas of 10 ha or greater, and where information was inconsistent it was upgraded or updated, while data gaps were plugged by information developed by Ontario Hydro staff. The environmental factors mapped for the study included:

- agriculture
- minerals

- recreational resources
- forestry
- biological resources
- human settlement/communities
- the cultural landscape
- heritage resources.

In addition, because an important consideration in locating transmission lines is the utilization of existing rights of way and linear severances, particular attention was paid to the mapping of existing utility and transportation rights of way and the administrative boundaries of local government.

The final task under scoping was the identification of environmental situations, defined as a concern or group of concerns that can be expressed on a map. These situations represent valued environmental components that must be addressed when considering the potential net impact of any new facilities. Ontario Hydro staff developed an initial listing of situations for each of the mapped factors. The final list was then reviewed by the two review groups.

Prediction

The review groups also reviewed and revised the environmental planning objectives proposed by Ontario Hydro staff. The planning objectives were developed to establish the relative importance of avoiding the various environmental situations determined during scoping. Each of the planning objectives was presented as a statement with the following components:

- a directive to avoid a particular environmental situation: e.g. avoid class 1 agricultural soils with good potential for specialty crops
- the value or importance of that situation (why it should be avoided)
- potential impact if that situation is not avoided
- mitigative measures that could be used to reduce the effects
- the 'net effect' or result if the mitigation measures are applied
- a 'constraint ranking' indicating the value of the resource and the net impact upon that resource.

The statements represented a form of impact prediction based on forecasting by an informed consensus of expert opinion. This approach to impact prediction lacks the scientific rigour of simulation modelling or statistical techniques, but it does offer an effective means to address issues of system interaction, dynamic change and uncertainty, while empowering stakeholders within the study process. This aspect was clearly evident in the case study, with the planner review group developing 37 environmental planning objectives and the interest review group listing 36 objectives.

Significance assessment

The first task in the determination of significance involved the ranking of the objectives developed by each of the review groups. Obviously, it would be impossible for the required transmission facilities to avoid all of the identified constraints. Thus, a mechanism to determine trade-offs was used which established the priority of the objectives within each factor and then considered all objectives in relation to each other.

An initial ranking was prepared by Ontario Hydro staff. First, criteria were established to rank the resource value of each objective. Then a second set of criteria was used to indicate the net impact of each of the objectives. The two sets of criteria were used to produce two rankings for each objective. These were then combined on an equal basis to derive a combined 'constraint rating'. The results of this process were reviewed, revised and finalized by the two review groups. The rankings were then employed in the preparation of the environmental constraint maps.

Environmental constraint maps are Ontario Hydro's primary tool for corridor identification (Smith and Cattrysse, 1987). The maps illustrate the distribution of the ranked objectives within the study area. In this instance, a 'five level constraint map' was used, corresponding to the division of constraint ratings into very high, high, medium, low and very low. To produce a constraint map, acetate overlays were used to isolate areas of lower constraint. The acetate overlay was then transferred to a 1 : 50 000 topographic sheet and the tentative corridor boundaries were adjusted to reflect the administrative boundaries of local government.

A constraint map was prepared for both the review group ratings. A third acetate overlay was produced by combining the corridor boundaries from each of these maps. In this manner, the resulting corridors reflected the concerns and priorities of both groups.

The alternative corridors were then checked by the two review groups and presented to local municipalities in meetings and to the general public through a series of information centres. The corridors were refined on the basis of the feedback and the study process was repeated at a scale of 1 : 10 000 to identify alternative routes.

The alternative routes reflected the integrity of the initial constraint rankings but also accommodated site-specific issues, such as the desire to avoid unnecessary severances. Thus, the initial ranking of environmental objectives was reassessed and modified to take into account the implications of the data collected at the 1 : 10 000 scale. In most cases, this meant the replacement of a more generalized objective with a more specific statement.

The end result of this process was the production of a composite map, illustrating site-specific constraint information for route identification. Alternative routes were then identified from the composite map by the two review groups, and a second round of municipal meetings and public open houses was conducted.

Evaluation

A number of alternative route segments were identified, evaluated and compared. The selected segments were then assembled into alternative routes (Fig. 8.3). At each stage, the comparison was based on the same criteria:

- *economic*: focusing on differences in the capital cost to construct the line
- *technical*: concerned with the engineering aspects of construction and issues of power system security and stability
- *environmental*: emphasizing the differences in potential impacts and net effects on the basis of the preceding mapping
- *socio-economic*: reflecting nature of displaced residents and businesses, community character, community facilities and effects on services.

The comparisons assumed that the 500 kV line would require a right of way 67 m wide, while the 230 kV line would require a 50 m wide right of way. The cumulative advantages and disadvantages of each route were then listed, and a preferred alignment was selected by Ontario Hydro. The potential effects of that route were identified, mitigative measures described and, following public review, any outstanding concerns or issues indicated. The final routes proposed by the utility are those shown in Fig. 8.3.

Conclusion

The Ontario Hydro case study is a good illustration of how impact assessment for linear facilities should be approached. The siting process to locate bulk transmission facilities west of London exhibited several strengths:

- The question of need was addressed.
- The objectives and characteristics of the project were clearly indicated.
- Public involvement and existing data bases were used to identify valued environmental components and the variations in their distribution were then mapped.
- Alternatives to the project and alternative routes were both addressed.
- Impacts were predicted on a consensual basis.
- Constraint ratings were developed to determine significance.
- Trade-offs were determined on a consensual basis by stakeholder review groups.
- A full range of alternative routes was considered.
- The mitigation of effects was a priority concern throughout the constraint-mapping exercise
- There was an extensive, multi-faceted and successful attempt to include all stakeholders in the study process.

However, one aspect not evident in the case study was the use of

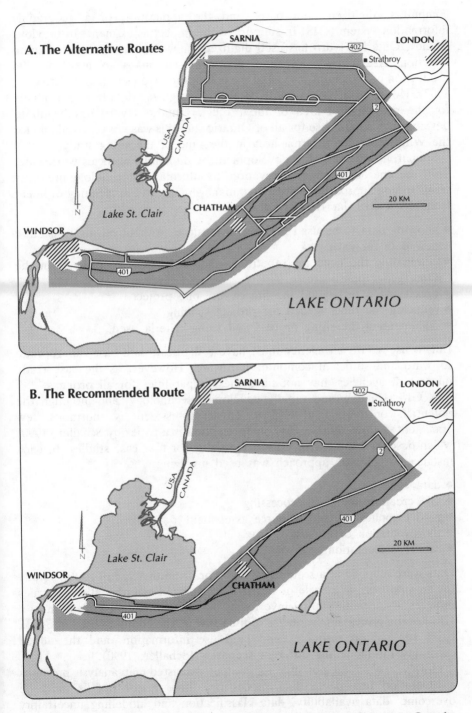

Fig. 8.3 Determining routes for bulk transmission lines in south-western Ontario

computers, digitized data bases and the application of a geographic information system (GIS) in support of the impact assessment. In previous studies, Ontario Hydro has used digitized data, a computer assisted route selection system (CARSS) and photo mosaic maps to assist in the production of constraint maps and the composite map (Smith and Cattrysse, 1987). However, a GIS can only be used if a digitized data base is available for the geographical area of the project. Ontario Hydro has begun to develop such a data base for all of Ontario, but this was not yet available for the Windsor–London–Sarnia area at the time of the siting study.

Similarly, GIS are under development by numerous agencies worldwide. They are one of the major innovations in information technology presently transforming approaches to resource management. In the planning of linear facilities, a GIS data base would assist by:

- allowing data to be used for several purposes without having to acquire new data for each task
- eliminating the need for repetitious data collection for projects in the same geographical area
- preserving data beyond the life of any one project
- reducing the time needed for project scoping
- documenting the siting process and siting criteria used.

But to use a GIS an agency must have a digitized data base, appropriate software and sufficient computer hardware. However, as the case study illustrated, agencies may not have these capabilities for all projects.

A GIS is a new and rapidly developing technology, and although the potential contribution of GIS to impact assessment is enormous, few documented case studies exist. One exception is a paper by Schaller (1990) which described the use of GIS in Germany for four case studies. In each instance, a common approach was used involving:

- data collection
- GIS-creation and data processing
- model application and resource assessment
- the evaluation of impacts
- results presentation.

Data were mapped at a scale of 1 : 5 000 or, as in the case of routing study for an autobahn, at a scale of 1 : 10 000. Digitized mapping at these scales requires a substantial initial investment but the subsequent utilization of a GIS has the principal advantages of time-saving, the ability to use large data bases for natural resources and land use information, and the added capability for modelling different scenarios (Schaller, 1990).

However, while the potential for computer-assisted map analysis promises a revolution for spatial data processing, three major pitfalls must be overcome: data availability, data classification and modelling uncertainty (Berry, 1987: 1409). In addition, other problems with GIS exist. Computer-assisted mapping and GIS overcome the practical difficulty of manually superimposing overlays for a large number of maps, but they do not necessarily circumvent other problems such as inaccuracy and the identification of cause and effect processes (Bailey, 1988). Inaccuracy can arise from differences in boundary definitions between different data sets and also occurs due to variances in the areal classification of data. In any mapping exercise, data are assumed to be evenly distributed within mapped cells. Rarely is this the case. Thus, it is difficult to map data precisely, especially at the boundary between one classification and the next.

A GIS may also generate other problems by giving a false perception about the quality of results. For example, composite maps give an empirical description of environmental factors but they do not consider the processes that control environmental responses nor the interactions between different systems (Bailey, 1988).

The real strength of a GIS is its ability to combine computer-assisted mapping with statistical analysis of spatial data. However, a major drawback is the problem of spatial autocorrelation. This situation arises when different data sets are superimposed on one another and a process relationship is assumed to link the data sets together, when in fact the correlation is simply there as a consequence of how the data have been mapped. How does the analyst distinguish true interdependencies from apparent dependencies arising from a spatial autocorrelation in data sets? Indeed, Moffat (1990: 215) has suggested that an

almost inevitable consequence of research into spatially located data is that the variates cannot be assumed to be independent. Often, interesting environmental problems concentrate on the interdependencies of two or more datasets. Serious inferential errors . . . can be made if the observations are autocorrelated and appropriately modified statistical techniques are not employed.

The dilemma here is that for a GIS to be used correctly in impact assessment it must, almost by definition, involve the use of sophisticated mathematical manipulations to control for errors from spatial autocorrelation. The use of such sophisticated methods detracts from the ease of understanding of the approach and, thus, defeats one of the principal advantages of its employment in impact assessment.

These problems illustrate a much larger issue for impact assessment at a strategic planning level. On the one hand, impact assessments for linear facilities must struggle for legitimacy and standing in comparison with technical, engineering analyses. This places a premium upon scientific rigour in impact assessment and prompts the demand for new technologies such as GIS to assist in the siting of facilities. On the other hand, pressures to make studies of impacts accessible to lay audiences, timely and cost effective, tend to counteract a resource manager's ability to use the full range of technical options that might be available. This situation brings to the forefront the basic question about the nature of impact assessment raised by this book. Is it a scientific tool for the analysis of environmental effects, or is it an approach to public decision making regarding development proposals?

The experience with linear facilities suggests that impact assessment can make a real contribution to sustainable resource management by continuing to evolve as means for public decision making. At a strategic planning level this requires:

- improving the scientific quality of impact assessments by explicitly distinguishing as separate activities the prediction of impacts, the measurement of their significance and the evaluation of their effects
- de-emphasizing the reliance upon compensation in project mitigation through the adoption of good planning practices
- supplanting the reliance upon professional judgement in impact assessments through the use of consensual methods of decision making.

CHAPTER 9

Waste management

Introduction

In many nations, waste management has emerged as a dominant environmental issue at an operational planning level. It is a problem that features a high degree of public attention to site-specific, community-based environmental protection issues focused on the practice of landfilling, alternatives to landfilling and the special problems associated with disposing of hazardous and radioactive wastes. It also is a problem that is continuing to grow in its severity and its impact. For example, the US Environmental Protection Agency (EPA) alone presently spends in excess of $1.4 billion annually on the problem of solid waste disposal (Dower, 1990).

Widespread attention to the problems of waste disposal dates from the 1978 recognition of the effects on local residents of the Love Canal site in Niagara Falls, New York. As both Gibbs (1982) and Levine (1982) have shown, there was tremendous bureaucratic and political resistance to acknowledging the extent and severity of the environmental and health impacts of Love Canal. The problems originated in the period from 1920 to 1953 when the site of an old shipping canal was used as a dump for municipal and chemical wastes. The principal company that used the site was the Hooker Chemical Company and in 1953 they finished filling the site, covered it with dirt and sold the land to the local board of education for $1. The sale also contained a clause releasing Hooker from any liability should any physical harm or death result from the buried wastes. A subdivision was built around the Love Canal site in 1955, with an elementary school erected over the dump site itself. Residents were unaware of the toxic chemicals buried beneath the subdivision. Despite complaints about health problems (including high rates of epilepsy and miscarriages) it was not until 1978, after an extensive series of media articles and exposure on the issue and the efforts of Lois Gibbs in forming a local residents' action group, that the problems of Love Canal received official recognition (Mazur, 1989).

159

Ultimately, the affected areas of the subdivision were evacuated and the site was sealed to prevent further outbreaks of pollution. However, while the site was decommissioned or remediated, it was not excavated, nor were all the *in situ* chemicals removed. The debate continues as to how safe the site remains.

The controversy surrounding Love Canal brought to the fore the whole question of landfilling, and in particular the problems of hazardous waste disposal. It also served to highlight that the field of environmental protection was changing. Environmental protection had previously meant pollution control and was usually applied to the control of emissions into the air and into water (Sandbach, 1982). Concomitantly, a substantial experience with impact assessment of air and water pollution also developed (e.g. Westman, 1985; Ortolano, 1984; Rau and Wooten, 1980), with some excellent case study analyses of air and water pollution management strategies. For example, Portney (1990) examined the development of air pollution control in the United States, Wood (1989) compared siting controls over air pollution sources in the United States and Britain, while Nishimura (1989) documented the Japanese experience with regulating air pollution. Similarly, Freeman (1990) has examined the development of water pollution policies in the United States, Mitchell (1990c) presented an overview of the international experience with integrated approaches to water management, while Colborn et al (1990) summarized the pollution problems of the Great Lakes, and Edwards and Regier (1990) and Caldwell (1988b) both addressed the application of an ecosystem approach to the management of the Great Lakes. But this experience is of limited utility in dealing with the questions posed for impact assessment by waste management.

Waste management involves the disposal of three distinct forms of waste: *municipal*, *hazardous* and *radioactive* (Maclaren, 1991). In brief, these can be differentiated as follows:

- Municipal solid waste consists of household wastes, commercial and institutional wastes, construction and demolition wastes, sewage sludge and incinerator ash.
- Hazardous wastes can originate in the household but most come from the industrial sector. They are defined as wastes which cause or have the potential to cause harm because they are toxic, corrosive, flammable, explosive, reactive or pathological.
- Categorized separately from other hazardous wastes because of their unique handling characteristics and slow decay rates, radioactive wastes are usually in one of two forms, high-level or low-level. Primarily spent fuel from nuclear reactors, high-level radioactive wastes generate heat and also require shielding to guard against their radioactivity. Low-level radioactive wastes include uranium mining and refining wastes and contaminated materials from reactor stations, institutions (e.g. universities

and hospitals) and industry. Low-level radioactive wastes do not generate significant amounts of heat nor do they need shielding.

Waste also differs from other resources in two major ways: (1) its resource base has been steadily increasing, and (2) the goal of waste management is 'to reduce rather than to increase or sustain the resource base' (Maclaren, 1991: 28). Thus, the problems of waste management necessitate consideration of how much waste is generated, how wastes should be disposed of and, most controversially, how and where waste disposal facilities should be located.

Waste management is a relatively new issue in environmental protection. It has come into vogue as a media topic and subsequently to the forefront of political agendas while impact assessment provisions were already in place. It is an issue that has arisen out of crisis, first as a result of public exposure to the problems at Love Canal and subsequently as communities began to realize that their capacity for waste generation was far exceeding their ability to dispose of that waste. Thus, waste management issues are usually accompanied by a sense of fear, perceived health risks and lack of trust in authorities. Environmental protection measures, including impact assessment, are seen by affected stakeholders has having failed to prevent or effectively mitigate the pollution problems they view as endemic to waste management. The result is further conflict and few instances of waste management have been free of controversy. This malaise is reflected in the rise in acronyms that characterize case studies in waste management, such as LULUs (locally unwanted land uses), NIMBY (the not in my backyard syndrome), and the latest, BANANA (build absolutely nothing anywhere near anybody).

Waste management represents a new challenge for environmental protection. Meeting that challenge requires a reconsideration of approaches to planning in the resolution of waste management problems and a re-evaluation of traditional applications of impact assessment to the problems of pollution control. In particular, waste management poses the following questions for impact assessment at an operational planning level:

- Problem identification:
  - How tractable is the problem?
  - How have issues of risk and uncertainty been assessed?
- Impact assessment:
  - What waste management facilities are required?
  - How should they be located?
  - What provisions are required for mitigation and monitoring?
- Interest representation:
  - Who are the affected stakeholders?
  - How should their concerns be addressed?

· How can issues of conflict and mistrust be resolved?

This chapter focuses upon the management of municipal and hazardous wastes, primarily in a North American context. First, it outlines the nature of the problems posed by municipal waste and discusses the general features of a strategy for waste management. The nature and scale of the hazardous waste problem in North America are then outlined. On this basis, the chapter turns to a discussion of the Superfund process in the United States to remediate, or decommission, old landfills. This is followed by a discussion of the various approaches to site selection for waste management facilities. The chapter concludes by analysing the implications of both these experiences for impact assessment at an operational planning level.

Municipal waste

In its simplest terms, the problem of municipal waste is one of an excessive amount of rubbish (garbage) and the increasing growth of rubbish with the advent of a disposable society. Since the early 1960s there has been a steady increase in the amount of use of disposable products: everything from nappies (diapers), food containers, razor-blades, and ball-point pens through to the planned obsolescence of appliances, such as televisions and machines, especially automotive products. The increased use of disposable products is further reflected in the fact that most of the increase in amounts of rubbish since 1960 has been in plastics and paper products (Maclaren, 1991).

Contemporary North American society has a dependency upon convenience which generates enormous amounts of waste:

- The average North American household produces 6.73 bags of rubbish in a week, or 29 bags every month. In a year that translates to 350 bags, with amount by volume equal to about 17 250 litres of rubbish that weigh in the order of 550 kg.
- Estimates vary, but by weight paper products comprise about 40 per cent of this total, food and yard waste another 30 per cent, with the remainder being metals (9 per cent), glass (8 per cent), plastics (7 per cent), wood (4 per cent) and toxics (1 per cent).
- Roughly 80 per cent of all rubbish ends up in sanitary landfills.

A relatively recent phenomenon, *landfills* emerged as the predominant solution to the growth in rubbish that accompanied the increase in technology and urbanization in the 1950s. Many advantages and disadvantages are associated with the location and operation of landfills (McKechnie et al, 1983). The advantages include:

- improvements in public health from sanitary disposal

- flexibility in operation
- an economical means of waste disposal
- properly designed and operated landfills with minimal impacts
- reclamation and use of submarginal lands e.g. pits and quarries
- rehabilitation of landfills after closure for use as recreational land, e.g. golf courses, parks etc.

These points are countered by the disadvantages, which include:

- Sites are often in differing administrative jurisdictions and political considerations often outweigh technical, economic and/or environmental factors in locating landfills.
- Landfills are often the target of public suspicion, fears and opposition, which can lead to sites being located in areas of least political resistance rather than in areas of optimum safety.
- Environmental protection measures (e.g. the prevention of leachate or the control of methane gas) may be expensive
- Landfills generate traffic and aesthetic problems for host localities.
- Landfill sites require sizeable amounts of land and compete with other requirements for scarce land resources.

The solution to the problems of landfilling was once thought to be incineration. During the energy crisis of the mid-1970s, there was a concerted push behind resource recovery through the generation of *energy from waste* (EFW). The basic strategy behind an EFW plant is to produce steam to generate electricity through the burning of municipal rubbish. The principal benefit of this strategy is a reduction of up to 90 per cent in volumes of rubbish destined for landfilling. However, problems associated with EFW plants have included:

- The costs of construction and of the energy produced have been uneconomic without substantial government subsidy.
- Toxic air pollutants (e.g. dioxin and furan) have been produced by the incineration of rubbish.
- The residual fly ash may have high levels of toxicity associated with it.

Thus, in the 1980s attention shifted towards various options for waste management that would reduce the need for landfilling and/or incineration. Primarily, this has meant the advocacy of the three Rs of waste management: *reduce, reuse* and *recycle* (Maclaren, 1991).

- *Reduce*: This entails source reduction and the generation of less waste. Reducing the amount of waste generated requires a major change in societal attitudes because of the dependency upon the disposable society and the convenience it provides. Some headway has been made in recent years through the imposition of new packaging regulations and the promotion of new consumer awareness programmes.

- *Reuse*: Reuse is commonly perceived to be the antithesis of growth and progress. Society is caught in the trap of the technological imperative: because the technology exists, people feel compelled to use it. Thus, people no longer reuse household products such as string, rubber-bands, razor-blades, old envelopes etc., which previous generations would use several times over. In North America today, reuse is associated with poverty and other social stigmas. Modern society is a product of an artificial, advertised image of the consumptive culture wherein conspicuous consumption is idealized rather than condemned. Without a broader societal change in attitudes, industry has been slow to adopt reuse strategies. For example, although the axiom 'pollution prevention pays' (Royston, 1979) has been adopted as corporate policy by the 3M company, it has not been widely embraced by other firms.
- *Recycle*: For most people and communities, recycling is the easiest starting point in a new strategy for waste management as it does not involve any significant lifestyle changes. Waste need not be reduced, nor consumptive patterns altered. Rather, recycling merely demands that the consumer and the community separate their rubbish into differing waste streams such as paper products, metals, glass and compost. The separated products are collected, sorted and then made available for recycling. Municipal recycling programmes have been widely adopted throughout North America over the past decade. They provide jobs, bring home the problem of waste management to the consumer and aid attempts at environmental protection by reducing the volume of waste left for disposal by incineration or landfilling. But not all products have proven amenable to recycling efforts. For example, whereas 50 per cent of all aluminium cans are recycled in North America, only 1 per cent of plastics are recycled. In other cases, recycling has fallen far short of its potential. For example, costs favour glass recycling but recycling capacity is inadequate to accommodate all of the glass presently being collected by municipal recycling programmes. Similarly, only 30 per cent of paper products are recycled because there are major limitations in the capacity of industry to process lower-grade paper products. In both instances, recycling presents the situation where the supply of raw materials exists but markets have not yet been found for them, rather than the traditional market mechanism of demand establishing the market.

Recovery has also been suggested as the fourth R of waste management. However, recovery is commonly associated with incineration and most waste management programmes stick to the three Rs to avoid the stigma of problems often perceived to accompany incineration. More generally, both incineration and landfilling are disposal options which rank below the three Rs in desirability. Indeed, reduction, reuse and recycling can be viewed as a hierarchy in a strategy of waste management designed to 'generate less waste, conserve more raw material resources, save more energy, and create fewer environmental impacts' (Maclaren, 1991: 35).

Hazardous waste

The health effects experienced at Love Canal typify public fears about the potential impacts of hazardous wastes. These range from skin burns and other irritations, to birth defects, neurological disorders and different forms of cancer.

In the United States, the legal definition of hazardous waste emphasizes their potential for increased mortality or irreversible and/or incapacitating illness. However, the regulatory approach to the management of hazardous wastes is not so much predicated on their definition 'as it is on the source of the wastes introduced into the environment' (Dower, 1990: 153). As a result, concern has focused upon the potential implications for environmental and human exposure that might result from the improper storage and disposal of hazardous wastes. This relates both to the risk posed by past disposal practices and to the imposition of guidelines on current disposal practices.

Although there are over 650 000 industrial generators of hazardous waste in the United States, it is estimated that over 95 per cent of the total amount of waste is produced by just 2 per cent of the producers (Dower, 1990). How industry presently disposes of this waste is a matter of some debate. A significant amount of hazardous wastes is treated, stored and disposed of by generator-owned facilities. However, according to estimates by the Congressional Budget Office (CBO), the most common means of hazardous waste disposal is through deep-well injection. This method utilizes old mines and salt domes to store wastes beneath the earth's surface, pumping the waste through thin, deep shafts. Together with landfilling, the CBO has estimated that nearly half of all hazardous wastes are disposed of in this manner. In contrast, both the EPA and the Chemical Manufacturers' Association (CMA) have estimated that the largest proportion of hazardous wastes is either stored in surface impoundments or discharged directly into surface waters (Dower, 1990).

The situation regarding past practices is even murkier. The EPA has estimated that there are over 27 000 abandoned hazardous waste sites in the United States, at least 2000 of which will require federal intervention if they are to be cleaned up. This is a conservative estimate and other federal agencies have put the total number of sites anywhere from 130 000 to 600 000, with up to 10 000 of those posing a significant environmental or health threat. As Dower (1990: 158) noted, 'If nothing else, these estimates reflect the range of uncertainty associated with past hazardous waste disposal; they do not hide the present magnitude of the problem.'

While Dower was discussing the situation in the United States, his comment has relevance to the status of hazardous waste management practices elsewhere in the world. Forester and Skinner (1987) presented the findings of an international working group on hazardous wastes, which reviewed hazardous waste management in Austria, Britain, Denmark, France, Germany, Italy, Japan, the Netherlands, Portugal, South Africa, Spain, Sweden and the United States. The working group summary concluded that many countries had effective legislation for the control of hazardous wastes. However, the experience among nations was highly variable and cross-national comparisons were complicated by a number of issues (Wilson and Forester, 1987). For example, there was a wide variation in how hazardous wastes are defined. Germany has followed the example of the United States and defines hazardous waste to separate it from industrial or municipal waste. In contrast, Britain has adopted a policy that favours the co-disposal of hazardous and municipal wastes in landfill sites, while Japan has a limited definition of hazardous waste which applies only to those wastes for which incineration or chemical treatment is particularly difficult. It is also difficult to compare quantities of hazardous waste produced because of the differences in definition and the absence of official statistics for countries such as Austria, Italy and Portugal.

The technologies for storage, treatment and disposal in the various countries ranged from temporary storage on the producer's premises, to incineration, landfilling and disposal at sea (Wilson and Forester, 1987). Incineration is extensive in Japan, Germany and France, less popular in Britain and the United States and infrequent in the other nations surveyed. Air pollution control was cited as the principal problem encountered with incineration.

Engineered landfills specifically designed for hazardous wastes predominate as the method of disposal in Denmark, Germany, Sweden and the United States. Britain is a notable exception to this philosophy. There co-disposal of hazardous wastes in municipal landfills is preferred. In a number of other countries, including Italy, Spain and South Africa, the general practice is for uncontrolled disposal of hazardous wastes by landfilling. In contrast, landfilling is prohibited in most circumstances in the Netherlands.

In assessing the siting of new facilities, the NIMBY syndrome was found to be a universal problem (Wilson and Forester, 1987). A variety of procedures has been adopted to adjudicate siting disputes. In Italy, local municipalities have a veto on siting, while Sweden, Denmark, Germany, the Netherlands and Britain all use some form of tribunal to review disputes.

Old or abandoned hazardous waste sites were just as big a problem in Denmark, the Netherlands and Germany as in the United States. Inventories have been conducted in all of these countries and also in France, Austria and Sweden, with extensive and systematic remediation programmes in effect in Denmark and the Netherlands. However, several countries (including Britain, Austria and Italy) had no specific provisions to fund the clean-up of abandoned sites.

Finally, the topic of hazardous waste management in developing countries was examined (Wilson and Forester, 1987). Most developing countries lacked adequate institutional arrangements for waste management. Most also lacked the necessary expertise to assess the potential problems posed by hazardous wastes. As a result, hazardous wastes are increasing throughout southern Africa, often as a direct result of lower disposal costs in less developed countries. Throughout Asia and the Pacific, a similar shortage of expertise was perceived but reliable data on hazardous wastes were lacking and a clear picture of the problem was not possible (Wilson and Forester, 1987).

Dealing with the past: Superfund and landfill remediation in the United States

The initial response in the United States to the problems at Love Canal was to assume that old landfills could be regulated under the 1976 Resource Conservation and Recovery Act (RCRA). This Act was introduced to provide cradle-to-grave regulation of hazardous wastes. It required the identification of wastes; the tracking of waste generation and transportation; and the establishment of standards for treatment, storage and disposal facilities. However, it was quickly apparent that the RCRA was inappropriate for the problems of abandoned sites as it was primarily aimed at the regulation of current wastes (Dower, 1990). Thus, the Comprehensive Environmental Response, Liability, and Compensation Act (CERCLA) of 1980 was passed. Better known as Superfund, the CERCLA established federal authority and mechanisms for the identification, prioritization, clean-up and emergency response to hazardous waste sites (Dienemann et al, 1991).

Both the RCRA and the CERCLA have been subject to revision. The RCRA was extensively revised in 1984 to replace its initial guidelines with specific requirements aimed at reducing the dependency upon landfilling of hazardous wastes and at closing loopholes in the original statute. The CERCLA was amended in 1986 by passage of the Superfund Amendments and Reauthorization Act or SARA, which revised and updated some of the fiscal and administrative aspects of the initial legislation (Dower, 1990). In conjunction, the two statutes provide the basis of the statutory and regulatory framework for waste management in the United States.

The Superfund law has two unique qualities: (1) it addresses past abuses rather than providing for environmental protection of the future; and (2) it authorizes the EPA both to regulate the management of hazardous wastes and to act as a contractor to conduct site remediation within those regulations (Dower, 1990). Superfund gives the EPA power to identify and enforce clean-up of abandoned waste sites. Sites are identified and placed on a priority ranking list for action on either an emergency, short-run or long-term basis. Identified sites are nominated by the EPA or by states for inclusion on a national priority list. Sites are ranked on the basis of EPA criteria which utilize a weighting scheme to reflect aspects such as toxicity and the potential for human exposure (Dower, 1990).

By the end of the 1980s, over 1100 sites had received a priority listing by the EPA under the Superfund designation, over half of which were in just seven states: New York, Michigan, New Jersey, Pennsylvania, Minnesota, California and Florida. The only state without a waste site on the national priority list was Nevada. However, the overall scale of the Superfund process is much larger, with over 30 000 sites identified for initial assessment. Actual remediation has been hard pressed to match the scope of the problem. As of 1989, only 48 sites had been cleaned up (Dower, 1990).

In addition to its enforcement provisions, the Superfund law also allows the EPA to establish liability for abandoned disposal sites. This means that individual firms can be held legally responsible for the costs of remediation at abandoned sites, even if they were not the sole disposers of waste at that site. This provision is intended to assist the EPA in avoiding protracted legal battles in the determination of liability. It also provides an incentive intended to foster private-sector initiation of site remediation.

However, progress in the implementation of Superfund has been slow (Bowman, 1988). Partly, this has been due to the high costs involved in site remediation. Moreover, because the overall budget of the EPA has not kept pace with the rising costs of Superfund, the requirements have had a serious effect upon other EPA programmes (Dower, 1990). Delays have also resulted from administrative and jurisdictional disputes (Lester, 1988). These have arisen as a consequence of poorly structured relationships between agencies, counterproductive decision rules and erratic implementation, which has tended to add more sites to the priority list faster than current sites can be remedied (Bowman, 1988). Finally, slow progress has often been the result of the complexities of the Superfund process itself and from site-specific disputes among stakeholders (Van Horn and Chilik, 1988; Marks and Susskind, 1988). The remediation efforts at the Lipari landfill site in New Jersey illustrate these problems (Dienemann et al, 1991).

The Lipari landfill case study

The Lipari landfill is located in Mantua Township, New Jersey, near the towns of Pitman and Glassboro (Fig. 9.1). The landfill is about 2.4 ha in size. The site is bordered by two streams and the surrounding land uses consist of orchards, public parks and a residential housing development (Dienemann et al, 1991).

The landfill contains municipal and household wastes, liquid and semi-solid chemical wastes and other industrial wastes. These wastes were dumped into a former sand and gravel pit to create the privately operated landfill during the period from 1958 to 1971, when disposal operations were halted. Although some steel drums were disposed of at the site, most of the wastes were not contained. Among the products in the landfill are solvents,

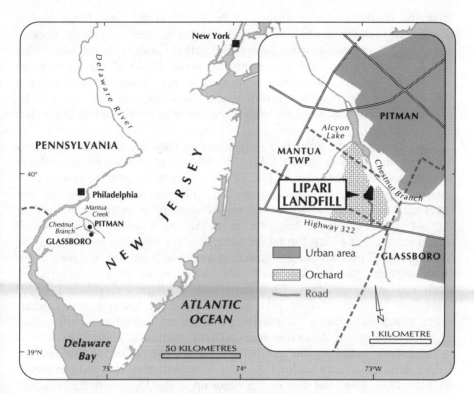

Fig. 9.1 Location of the Lipari landfill in New Jersey

paint thinners, formaldehyde, paints, phenol and amine wastes and resins (Dienemann et al, 1991). Estimates indicate that upwards of 11 million litres of chemical and industrial liquid wastes were placed into the landfill over its lifetime. Much of the present concern focuses upon the subsequent migration and dilution of this liquid waste as leachate into the local water table.

The effects of the landfill on the adjacent streams and a local lake were the principal factor in its closure in 1971. At that time, a multicoloured, foul-smelling leachate was observed to be seeping out of the landfill. These observations were supplemented by complaints from local residents, who apart from the odour were suffering from headaches and nausea.

In 1972, the New Jersey Department of Environmental Protection charged the operators with violating the state water quality legislation. The operators responded by digging drainage ditches to intercept the leachate and by spreading lime and soil to suppress the odours. Neither action was effective and a court order was issued in 1974 requiring that the site be cleaned up.

However, no further action was taken until 1978, when the EPA first sponsored and then conducted a series of monitoring studies on the leachate from the landfill under provisions of the Clean Water Act. These studies discovered that the leachate contained significant levels of volatile organic compounds, suspected carcinogens and other priority pollutants. This finding prompted the EPA to commit just under $800 000 over three years to clean up the site. This money supported further studies which delineated the extent of the pollution problem and recommended that the site be capped with clay and encircled 'with a cementitious slurry wall' (Dienemann et al, 1991: 20).

The passage of the Superfund legislation in 1980 extended the necessary authority and funds to the EPA for it to investigate and remediate problem landfills such as that at Lipari. The EPA could now become the lead agency for the Lipari site. In that capacity, the EPA conducted two further site investigations which led to formal remediation plans being formalized in August 1982. The plan called for a 76 cm thick wall around the site, which would intersect the clay formation lying beneath both the landfill and the local aquifer. This containment structure would be supplemented by a 1 mm thick synthetic cap over the whole site, which would then be covered by fresh soil and vegetation. Consistent with guidelines established under Superfund, the remediation plan also provided for leachate and ground-water treatment to ensure a cost-effective mitigation and 'adequate protection of, public health and the environment' (Dienemann et al, 1991: 20).

The Lipari landfill was the number one ranked site out of 546 listed on the EPA's national priority list for 1983. The containment system was completed by June 1984. The total cost of the clean up at the Lipari site had reached $4.4 million at that juncture. However, the containment measures were not the end of the problems at the Lipari landfill. In fact the controversy was just about to begin in earnest.

In the summer of 1985 a residents' association entitled PALLCA (the Pitman, Alcyon Lake, Lipari Landfill Community Association) began to lobby for complete removal of the contaminants. Consistent with revised regulations introduced in 1985, the EPA was by then authorized to address the remediation of contained contaminants. The EPA considered four possible alternatives:

- No action.
- Enhanced containment, involving the pumping of the aquifer below the landfill at a cost of nearly $7 million over 25 years.
- A procedure known as batch flushing of the containment system, whereby the landfill would be drained of its contaminants, flushed several times with fresh water to remove residual contaminants and the resultant leachate treated. Batch flushing would take 15 years and cost about $9 million.
- Complete excavation and off-site disposal of the contaminated material, estimated to cost $290 million.

The EPA chose the batch flushing option. It was cheaper than complete excavation, while the first two options were rejected as they could not provide public health and the environmental protection within the necessary time frame. The EPA choice was controversial. PALLCA, the US Office of Technology and several prominent politicians were critical both of the approach and of the decision by the EPA in the Lipari clean up. Their consensus was that the EPA 'could have and should have selected a more aggressive cleanup for the Nation's #1 Superfund site' (Dienemann et al, 1991: 22).

Flushing was criticized as being unproven and less effective in treating contaminants than incineration. The debate over its effectiveness illustrated that a fundamental issue within the Superfund process was the determination of what was meant by 'clean'. For the EPA, flushing would remove up to 90 per cent of the contained contaminants, providing a clean up of the Lipari site. For residents and opponents of the flushing approach, however, anything short of complete removal of the contaminants would not constitute an effective clean-up.

The controversy over the clean up was featured extensively in the media, and communications between the various key stakeholders became strained. The conflict was further complicated by delays due to litigation between the EPA and the various principal parties held liable under Superfund provisions for the costs incurred. However, the delay in implementing the clean-up did permit a series of health studies to be conducted. These showed that there were no causal links between the landfill and health problems of local residents. These findings acted to calm the dispute somewhat. Relations were further improved when the EPA responded to the concerns of the PALLCA by amending its remediation plan to reduce the time-span for implementation to 10 years, commencing in 1992. The agency also adopted a more aggressive stance in the remediation of off-site areas (Dienemann et al, 1991).

The Lipari landfill case study illustrates the lengthy time period involved in site remediation. Over 20 years have elapsed since the discovery of significant problems with the site. And while the pollution risk may have been contained, it will be at least another decade before the site is fully remediated. Not only is the process long, but it is also very expensive. Total costs for the Lipari landfill remediation will exceed $40 million. With over 1100 sites on the priority list already and the prospect of over 30 000 more potential candidates for remediation, the cost implications of Superfund are enormous. Complicating the process are the site-specific nuances of each landfill. This limits the ability of the EPA to utilize a standardized approach to remediation, as the geographical, technical and political circumstances of each site necessitate an individual plan of action. As Dower (1990: 187) concluded: 'Other alternatives for reducing the legal and transaction costs associated with the current Superfund program should be investigated, for these costs stand as a significant barrier to efficient and effective implementation of a national clean-up program.'

Planning for the future: the facility siting dilemma

Waste facilities, especially landfills and incinerators, have the potential for significant impacts on both the biophysical and the socio-economic environment. These effects can include both air and water pollution, increased traffic congestion, lower property values, health risks and habitat loss. Ever since the problems at Love Canal burst into public prominence, waste facility siting has become a steadily more controversial and bitter process, often characterized by conflict, antagonism and deep mistrust between stakeholders.

The conventional approach to waste facility siting is similar to that for linear facilities described in Chapter 8. Typically, a conventional siting process is phased and involves:

- *Defining a candidate area*: this is usually based on criteria such as the location of waste sources, volumes of waste produced, transportation limitations and political boundaries.
- *Area screening*: using the process of constraint mapping, potential sites are identified. Typical criteria for landfills would involve the mapping of hydrogeological conditions, soils and land-use factors.
- *Site evaluation and selection*: the candidate sites are compared on the basis of technical criteria, which are weighted, scored and evaluated. A preferred site is then identified and proposed for approval.

Criticism of this process centres on its emphasis upon technical criteria (at the exclusion of social concerns such as psychological stress, perceived health risks and community fears based upon scientific uncertainties) and the minimal role given to the public in decision making on site location, evaluation and selection (Maclaren, 1991).

In contrast to this product-oriented approach to siting, a process-oriented approach has been suggested which emphasizes community co-operation rather than technical criteria in determining site location. Under this alternative approach, candidate areas are identified by communities volunteering to host the required facilities. The potential host communities are then screened according to the technical criteria. Willing communities that have a suitable site are thus identified and the siting process centres on the selection of the host community among those that are suitable. The first successful implementation of the volunteer approach in North America was in Alberta and involved the location of an integrated hazardous waste treatment and disposal facility near the community of Swan Hills (McQuaid-Cook and Simpson, 1986).

The Swan Hills case study

In 1984, the Alberta government announced that a provincial hazardous waste treatment and disposal facility would be built 20 km north-east of the community of Swan Hills (Fig. 9.2). The site was environmentally sound.

172

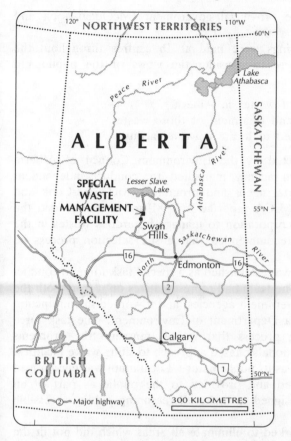

Fig. 9.2 Location of the Swan Hills waste treatment facility in Alberta

The area is one of glacial deposits and boreal forest, and the site itself lies within a large tract of provincially owned land with no neighbouring land-uses. More significantly, the decision to build at Swan Hills marked the first time in North America that social acceptability had been used as a primary criterion in determining site location for a hazardous waste facility.

The announcement marked the culmination of a three-year site selection process. Planning for the facility had been initiated in response to Alberta's need for an integrated waste management facility to handle the 100 000 tonnes of industrial and hazardous waste produced each year in the province. In 1979, a moratorium had been placed on all off-site waste disposal facilities in the province after adverse public reactions to a private-sector proposal for a treatment plant near Edmonton. In addition to imposing the moratorium, the government initiated a committee to review the need for a provincial waste management plan. The committee concluded that Albertans were obliged to develop a waste management plan, that a waste treatment system should be developed, and that a leadership role

173

should be assumed by the provincial government (McQuaid-Cook and Simpson, 1986).

A series of public hearings were held at 16 centres throughout the province. The hearings were to consider the views of the public and determine:

- the types and amounts of wastes in Alberta
- the options for storage and treatment of those wastes
- criteria that could be used to locate a waste facility.

The hearings were conducted by the Environment Council of Alberta (ECA), which is a Crown agency mandated to 'conduct independent assessments of environmental needs and programs' (McQuaid-Cook and Simpson, 1986: 1031). The ECA reported in 1981 and recommended the establishment of a Crown corporation to manage hazardous wastes in the province. The ECA also recommended that a site selection process be initiated.

A hazardous waste team was established as a special task force to conduct the site selection process. The team had representatives on it from both the public and from various government agencies, and it worked in conjunction with staff from the Alberta Department of Environment. The task force decided to adopt a siting process that would transcend the problems characteristic of the conventional, phased approach to siting waste facilities. The basis of the approach was its public nature. Local input was encouraged and all data were displayed and provided to the public as part of an information programme designed to make all reasoning public, accessible and comprehensible.

The selection process worked to eliminate all areas which did not fit the design criteria. A constraint mapping exercise was used to screen the whole province at a scale of 1 : 1 000 000, followed by regional assessments at a scale of 1 : 250 000. Overlay mapping was utilized and the siting criteria used to determine constraints included physical, biological, land-use and human factors. However, what was distinctive about the Alberta process was not this phased, physical assessment, but the public participation that accompanied it.

Throughout 1981, every municipal and local government administration in the province received a briefing on the siting process. This included a display of the provincial map and an outline of the general siting criteria necessary for the environmentally safe location of a waste facility. In this manner, over 120 community meetings were conducted in the province. At their conclusion, local authorities were given two options. They could either opt to go no further in the siting process, or they could request that their area be included in detailed regional analyses. This request had to be in the form of a written invitation, signed by the local administrator. Invitations to assess the suitability for a site were received from 52 out of a possible 70 jurisdictions in the province (McQuaid-Cook and Simpson, 1986).

The next phase of the physical assessment involved regional analyses for those areas that had chosen to proceed with the siting process. All studies were presented at open public meetings, at which juncture the basic requirements for an integrated facility for waste treatment and control were also outlined. Where there was public opposition, areas were dropped from further consideration. Environmental constraints caused other areas to be dropped. Twelve of the studies areas requested further consideration as a possible host community for the waste treatment facility and each identified three possible locations for further, detailed assessment.

Those that chose to proceed tended to view the facility relative to its potential benefits (McQuaid-Cook and Simpson, 1986). In particular, these were perceived to include such local benefits as employment prospects, an improved tax base, financial benefits from visitors, improved services, opportunities for industrial diversification and utilities upgrading.

Detailed screening ruled out seven of the communities. The remaining five communities each then requested that drilling be conducted to confirm the suitability of the local geology. These five communities then received a series of three seminars to clarify for the local public the programme background, the rationale for the facility, the technology involved and the characteristics of the proposed plant. Public support was a precondition of site location and plebiscites were held in each of the five communities to gauge this. Voter participation was high, and votes in favour of the facility were 77 and 79 per cent for the two primary candidates (Swan Hills and Ryley). The final decision was made by the provincial Cabinet and with 'a strong indication of public acceptance, and proven environmental suitability, Cabinet chose the site near Swan Hills for development of the hazardous waste treatment plant' (McQuaid-Cook and Simpson, 1986: 1034). Construction of the facility began shortly after the decision had been announced. The Alberta Hazardous Waste Management Facility has been operating since September 1987.

The Alberta process was remarkable for a number of reasons. Most significantly, the process took only three years to locate a hazardous waste treatment facility. The brevity of this time-line contrasts sharply to the situation in Ontario, where a traditional phased approach required over 10 years to identify and approve a site. The quick decision period reflects public acceptance both of the need for the facility and of the planning process used in its location. The Alberta process reduced conflict and allowed communities to decide for themselves if they wished to host the facility. Voluntarism meant that there was no imposition of a LULU or opportunity for a NIMBY response to occur. If there was opposition to the proposal in a community, it was dropped from consideration. One way in which public opposition was diffused was the decision to make all information readily available to the public throughout the siting process. The general public were not condescended to by the project planning team, nor were they made to feel silly if they had fears or concerns. Consequently, the facility was not viewed negatively as a LULU and there was a general perception of the facility as beneficial to the host community.

The principal criticism of the approach is that voluntarism preys on the economic vulnerability of remote communities. This argument suggests that a noxious facility will be disproportionately attractive to communities that are more economically vulnerable and/or deficient in the services seen to accompany site location such as improved transportation links and utility services. A second equity issue relates to the question of who decides whether or not a community should volunteer to host a facility, i.e. how is local consensus derived, and what happens if there are no willing communities? Both these concerns are legitimate but they are not insurmountable. Certainly, in the Alberta situation, little evidence indicates any discontent with either the process or the final decision to build the facility at Swan Hills.

Implications for impact assessment

Taken in conjunction, the experiences with Superfund remediation and the facility siting dilemma underscore some of the fundamental issues for reform of impact assessment procedures. Waste management poses questions that are very real and tangible to local communities. Unlike other environmental problems that appear remote to everyday life (such as ozone depletion, climatic change or acid rain), waste issues are easily recognizable within the household, the workplace and the community. Therefore, public awareness is high and there is a predisposition in most communities for opposition to proposals related to both municipal and hazardous waste management. At the same time, familiarity does not obviate waste issues from being scientifically complex and fraught with uncertainty. Thus, administrators and planners are still faced by problems of incomplete, contradictory and absent data. They must still find ways to balance conflicting interests and priorities. And, they are still in need of information to assist with decision making.

Many have assumed that the advent of impact assessment would circumvent these problems. Indeed, it can be argued that impact assessment was conceived with exactly these problems in mind. However, as stated earlier, most waste management problems have emerged in the period since impact assessment came into being. Yet problems persist and in some cases appear to be getting worse. What are the implications for current impact assessment practice at an operational level from the experiences with waste management?

Problem identification

Issue attention is a key determinant of how problems are approached in resource management. Waste issues have been at the forefront of policy agendas in many communities over the past decade and one consistent element in all controversies appears to be media attention. Has the media been playing a role as a vehicle for public information? Or, has the media acted more as a policy participant, functioning as an active stakeholder, advocating particular strategies for environmental protection?

The *role of the media* in waste management issues was the focus of a study by Mazur (1989), who examined the role of the media in communicating risk using Love Canal as a case study. He concluded that media coverage always aids protestors in environmental cases whether or not their case has merit. Moreover, his findings indicated that protestors and journalists have a symbiotic relationship. Protestors supply a focus that sells news, while media attention is an irresistible and intoxicating enticement to most protestors. Protestors are usually a minority but constant media attention tends to give the impression that their views are more widely shared. At the same time, Mazur found that the amount of media coverage was positively correlated with the level of public opposition: the more media attention there was, the more public opposition there was, irrespective of whether the media coverage was for or against the specific proposal.

Results from research on an EFW plant in London, Ontario tend to confirm these findings. Given (1989) monitored the concerns of residents in response to an EFW facility built by a local hospital. She found that most people got their information on the facility and its potential effects from the local media. Subsequent research using content analysis revealed that the principal source of information for the media was environmental groups and activists opposed to the plant (Smith and Given, 1989).

However, these findings are by no means conclusive. Indeed, Portney (1991) has reported results which did not support the idea that public opposition is correlated with the amount of media attention an issue receives. At the same time, weaknesses in his own data prevented Portney from dismissing the proposition entirely. Such discrepancies are typical of research in its infancy, and are indicative of the fact that the role of the media in environmental issues has not been extensively researched. Studies that have been done (Bendix and Liebler, 1991; Sandman et al, 1987; Rubin and Sachs, 1973) have tended to indicate that the media can play a large role in policy making. But, at the same time, these studies concur with Mazur (1989) that the specific content of media reporting in environmental issues is often superficial and is more concerned with sensationalism than scientific precision. As a consequence, public perception of environmental problems is often shaped by a sense of fear and distrust carried by media extrapolations from other events that may or may not fit present circumstances. This fear in turn acts as a integral antecedent for conflict.

Fear is closely linked with people's *perception of risk*. It is this sense of risk that underlies the attitudes of stakeholders towards problem tractability. A central concern in waste management is the question of how to deal with issues of scientific uncertainty. In most circumstances, the effects of long-term exposure to potential hazards are simply unknown. Meanwhile the scale of the waste problem appears to be growing far faster than the capability to remediate those sites already identified. The problem faced by resource managers is that fear, risk, uncertainty, inadequate data and the increasing awareness of problems have tended to coalesce as a seemingly insurmountable obstacle to effective waste management leading to the *NIMBY syndrome*.

Hallman and Wandersman (1989) have suggested that much of the NIMBY syndrome is correlated with perceived risk. They then present a good overview of the extensive literature on risk and how it is perceived. Perceived risk is a function of risk severity and the amount of fear associated with the risk. Voluntarism does not affect the perception of risk, but it does affect the acceptability of that risk. Lastly, the literature suggests that public perception of risk is affected by cultural beliefs, the plausibility of the threat, its proximity, previous experiences, the presence of children and prior or current health problems (Hallman and Wandersman, 1989).

The risk associated with waste management is largely related to the perceived impact on health. Andrews (1990) has noted that quantitative health risk assessment has been refined into an analytical procedure involving:

- *hazard identification*: the determination of a health risk from a substance
- *dose–response assessment*: the linking of the amount of exposure to the degree of toxic effect
- *exposure assessment*: the estimation of how many people are affected and to what degree
- *risk characterization*: a summary of the overall health risk to the population.

However, this sense of certitude is illusory, for as Andrews (1990: 179) also stressed, 'data remain scarce and expensive, basic mechanisms (let alone magnitudes) of toxicity remain uncertain, and exposure patterns and compounding factors will always be too complex to identify with certainty.'

The nature and sources of uncertainty in risk assessment have been discussed by Suter et al (1987). Of most concern to waste management are 'defined analytical' uncertainties, which result from modelling errors, natural variations in predictability or errors in measuring variables. The result of these uncertainties is that the level or frequency of risk is incorrectly determined and the scientific basis for judgements becomes suspect (Suter et al, 1987). In the realm of research, such variations are the basis for debate, discussion and scientific symposia. However, in the cauldron of public policy, they can be the death-knell for a proposed facility or remediation plan (as was nearly the case for the proposal to flush the Lipari landfill).

Because risk analysis is not an exact science, but instead relies upon estimates of probability, uncertainty is an inevitable, 'intrinsic characteristic of risk analysis' (Roberts and Hayns, 1989: 491). Consequently, the public examination of quantitative risk analyses tends to be severely constrained by the adversarial nature of most regulatory proceedings. Instead, other approaches to the evaluation of risk and the assessment of uncertainty have been advocated to circumvent these problems. Roberts and Hayns (1989, 493) stressed the 'need for mechanisms of peer review to identify the critical areas of judgement that enter into a risk analysis, preferably before the public stages of a formal inquiry.' In Britain, one such mechanism has been the use of formal advisory committees to give expert advice to the government on environmental matters. The role of formal advisory committees has been evaluated by Everest (1990). His analysis suggests that advisory committees evolve from a position of providing professional scientific advice to a position of providing 'trans-scientific' advice, which includes aspects of public awareness and judgement. Formal committees also offer protection against political emasculation, which is the chief caveat restricting the utility of advisory bodies in environmental policy making (Smith, 1982a, b).

Institutional arrangements

Two recent studies have addressed the question of the institutional arrangements for waste management in the United States. Davis and Lester (1988) used the policy implementation framework of Sabatier and Mazmanian (1981) (Fig. 3.3) to examine hazardous waste policy in the United States, while Bowman (1988) used the same framework to examine Superfund implementation. The studies were critical of the institutional arrangements for providing the EPA with a mandate that necessitates a broad consensus among stakeholders for implementation. At the same time, the EPA cannot mount remedial actions without a formal agreement with affected state governments. It is not surprising, therefore, that delays have characterized the Superfund process, and the 1986 introduction of SARA can be seen as an attempt to compensate for this initial deficiency. However, with the scale of the economic resources required for implementation, it is likely that Superfund will continue to be bogged down in an administrative quagmire.

The Superfund example illustrates how good legislation is but one component within an effective institutional arrangement and that the leverage exerted by the processes and mechanisms for implementation can be decisive. This was also found to be true in the example of Swan Hills, which illustrates that the existing institutional arrangements are often adequate for effective siting, provided appropriate processes and mechanisms are utilized. Within impact assessment, it would appear that an appropriate process for waste management is one that makes explicit provision for public participation mechanisms.

Impact assessment

The central issues facing impact assessment for waste management relate to:

● the determination of what facilities and/or activities are required

- facility siting and location
- the mitigation of impacts
- the monitoring of effects.

As always, effective impact assessment at an operational level starts with scoping. For waste management issues, scoping increasingly entails the consideration of an integrated waste management strategy, incorporating the three Rs (reduction, reuse and recycling). Most scoping exercises have paid only token attention to the three Rs in the justification of a proposed control, treatment and disposal facility. But, within the next decade, it can be expected that impact assessments will begin to adjust to the ethics of sustainability and consider strategies for integrated waste management more seriously.

The move to planning waste management strategies will necessitate that the weighting of technical and scientific information over social acceptability will also have to undergo a transformation. Particularly at the evaluative phase of impact assessment, *social acceptability* can be expected to increase in significance as a central criterion for waste management projects. The adoption of waste management strategies will entail a shift in values away from the technological imperative. Concern for health and environmental protection will take precedence as communities become increasingly risk adverse. This change in perspective will translate into criteria for social acceptability underpinning the siting process, rather than the traditional, phased approach to technocratic problem solving. Waste facility siting will become a more overtly political process, as was the case in Swan Hills, than an exercise in rational reductionism as it has been under the traditional approach of decide, announce and defend.

This is not to suggest that a paradigm of societal learning can solve all the dilemmas inherent in waste facility siting. Mechanisms still have to be developed to incorporate the consideration of cumulative effects, and provisions for monitoring have to be greatly strengthened. The uncertainty surrounding hazardous wastes dictates that more attention must be paid to the potential for cumulative effects from toxic substances. The Superfund experience thus far indicates that sealing old dumps may be viewed as an appropriate means of achieving safety in environmental protection, but there is tremendous pressure for site remediation to clean the environment by removing all *in situ* pollutants. No truly effective determination of 'clean' is possible without greatly increased attention to the potential for cumulative effects. The increased use of *environmental audits* is one initiative which is likely to spur increased awareness of cumulative impacts, while at the same time underscoring the value of a well-conceived programme of monitoring.

Impact assessment *monitoring* has not been extensively developed. Indeed, in many jurisdictions, there are no legislative provisions for monitoring within impact assessment. Certainly within waste management there are significant benefits that could be derived from initiating effects monitoring and public concerns monitoring at the scoping stage and surveillance monitoring to accompany the implementation and operation of a waste management strategy:

- Effects monitoring would assist in the determination of risk and the evaluation of alternative management options.
- Public concerns monitoring would enable the incorporation of the political dimension of the siting process with the phased analysis of environmental suitability.
- Surveillance monitoring would assist in screening for cumulative effects and the ultimate public assurance of risk reduction.

Collectively, these concerns emphasize the importance of the evaluation, mitigation and monitoring stages of impact assessment at an operational level. However, as the examples have shown, the real issue is not just the elements of the impact assessment process itself but the incorporation of stakeholders into that planning process.

Interest representation

Waste management provides a graphic illustration of the principal elements predetermining a NIMBY response to development proposals:

- real and perceived inequities in risk sharing
- a lack of trust in regulatory decision making
- the desire for real community self-determination.

Not surprisingly, NIMBY reactions and the nature of the NIMBY syndrome itself are topics of great interest to researchers in the waste management field. In one of the more comprehensive, comparative analyses, Finsterbusch (1989) reviewed 25 case studies in the United States to assess community responses to hazardous wastes. A common pattern was discovered:

- Predisposal was characterized by poor institutional arrangements for waste management.
- Disposal of wastes was not sound but polluters were unaware of the health risks they might be creating.
- The discovery of problems left affected stakeholders confused and frustrated by official responses that were not responsive to their concerns. Only in two cases did the media play a major role in publicizing risks.
- A reactive phase followed with affected groups organizing and becoming more sophisticated in their protests. Meanwhile, dumping continued for an average of 4.1 years after initial discovery of problems.
- The resolution of the problem usually involved removal of the danger but little actual removal of toxic substances.

The 25 cases also indicated that the efforts of particular individuals (such as Lois Gibbs in the Love Canal instance) were crucial in stopping pollution and obtaining redress. In addition, Finsterbusch (1989) suggested that the Love Canal example revealed:

- that the public tends to be passive and dependent
- the importance of a leader
- that the media can help foster accountability
- that science can be politicized
- that conflicts between officials and the public are inevitable.

Quarantelli (1989) examined the characteristics of groups formed in response to hazardous waste sites. Typically such groups have less than 100 members. Formal membership lists are rare and the group is largely driven by a small and very active core. Participation is always a part-time activity and the core members were usually those who joined at the initial phase of the group. Most groups had a disproportionate number of women as members. However, the implications of gender were not expanded upon and they have not, generally, been assessed by other studies.

One exception was the work of Spain (1986) who did offer some preliminary thoughts on this topic. For example, it is possible that the higher participation of women and the leadership role they have taken in waste management issues can be related to the perceived health threat of hazardous wastes. Children are the most vulnerable to health effects of toxic substances. The traditional role of women as primary care-givers to children would thus result in them being more sensitized to, and aware of, potential impacts from poor waste management practices. A second, potentially more significant, implication of gender might relate to models of power and how disputes are resolved. Most decision making structures have traditionally been male-dominated. It is no accident, therefore, that the dominant means of dispute resolution in such forums are closely allied to adversarial notions of power which are innately male in concept. In contrast, feminist models of power relationships are rarely based on adversarial concepts. This incongruence may offer one explanation for the conflict other researchers have considered inherent within waste management controversies and also offers one explanation for why consensus-based strategies for siting (such as at Swan Hills) hold such promise for the reform of dispute resolution. Thus, further research on power relationships among and between differing stakeholders that includes some consideration of gender is needed.

Portney (1991) examined the NIMBY syndrome and how it affects the siting of hazardous waste treatment facilities. He suggested that the process of facility siting must be viewed as much more explicitly political than in the past. Patterns of opposition to facilities are an extension of broader political, social and psychological attitudes towards risk and the environment. Thus, siting processes will only be successful if they conform to those attitudes. Conversely, conventional, phased and technical approaches to locating facilities that ask stakeholders to alter their attitudes on the basis of site assessment studies will be immersed in conflicts (Portney, 1991).

In a similar vein, Mazmanian and Morell (1990) have suggested that the NIMBY syndrome reflects a 'failure in democratic discourse'. They traced the origins of NIMBY to the inherent imbalance in the distribution of costs and benefits associated with any project. But, as they pointed out, NIMBY extends beyond narrow self-interest and reflects a broader distrust of science and technology. A fear of health risks and a general lack of faith in both regulatory agencies and processes compound this factor. Furthermore, if the NIMBY syndrome is to be circumvented, 'these causal factors will have to be effectively addressed' (Mazmanian and Morell, 1990: 127). The solution to NIMBY lies with a number of factors:

- how the need for a project is defined and by whom
- the characteristics of the project
- the standards of fairness evident in the siting process
- the nature of the siting process itself
- provisions for long-term monitoring of public safety.

Collectively, these studies highlight the need for impact assessment to take interest representation into explicit account in the design of procedures. The waste management experience is testimony to the fact that the days of decide, announce and defend are past. The inclusion of a public information programme or some token attempt at public involvement is no longer sufficient as a strategy for interest representation. Impact assessment must account for the stakeholders in any issue, accommodate their goals and utilize approaches that empower communities within decision making. Stakeholders must be given a role in defining not just the substantive content of planning but the approach to planning itself.

However, the desire to avoid NIMBY and other undesirable responses is one thing; having the means to do so is another. Not all jurisdictions or situations will be suited or amenable to full voluntarism as practised in the Swan Hills example. One commonly suggested solution to this problem is the invocation of alternative techniques for environmental dispute resolution (EDR). But, as was emphasized in Chapter 4, the application of EDR in a pro-active manner is constrained by a number of limitations and caveats. Again, this appears to be another area where further research is needed if the goals of increased interest representation in impact assessment are to be realized.

Conclusion

The experience with waste management indicates a number of areas where impact assessment can be strengthened at an operational planning level. The approach to impact assessment itself needs to become more overtly political. Moreover, as this occurs, the process of planning must be adjusted throughout the various stages of impact assessment, with particular attention to the need to revise the approaches used:

- in scoping
- in the weighting of criteria at the evaluation stage
- in the requirements for monitoring of effects, compliance and public concerns.

One possible means of achieving these aims is to build on the strengths of the experiences to date. Thus, it is suggested that impact assessment at an operational level could be improved through the adoption of provisions requiring:

- That a task force of stakeholders be initiated at the scoping phase of an impact assessment to determine the nature of the problem to be solved and the planning approach to be utilized. The task force would then act as the facilitator for interest representation throughout the planning exercise. This would include responsibility for co-ordinating approaches to stakeholder participation in studies and the incorporation of social acceptability criteria with criteria for environmental acceptability.
- Formal guidelines for the impact assessment process be available to provide direction to proponents in differing resource sectors as to the expectations for planning through impact assessment. For example, guidelines for the assessment of waste management strategies would:
  · include the incorporation of a more overtly political approach to planning
  · specify the consideration of cumulative effects
  · require the establishment of a monitoring programme for effects, compliance and public concerns from the scoping stage through to plan implementation and facility operation.

CHAPTER 10

Implications for sustainable resource management

Impact assessment at the crossroads

Impact assessment continues to struggle for legitimacy and standing in comparison with technical, economic and engineering analyses. This quest has placed a premium upon scientific rigour and a focus on environmental impact statements attempting to outline the character of project effects. At the same time, there are growing pressures to make studies of impacts timely and cost-effective, as well as accessible to lay audiences. It is a combination that tends to counteract the ability of resource managers to use the full range of technical options that might be available. Also there are limitations on the ability of science to delineate project effects effectively and accurately due to data limitations and the uncertainty that characterizes many developments. This situation brings to the forefront the basic question raised in this book. Is impact assessment a scientific tool for the analysis of environmental effects, or is it an approach to public decision making regarding development proposals?

Storey (1986) suggested that the focus upon impact statements has placed the emphasis on the identification and description of likely outcomes of particular projects. Indeed, most impact assessments have not considered impact prediction, significance assessment and evaluation as separate, discrete elements in the determination of effects. Overall, impact assessment has exhibited an over-reliance upon professional judgement in lieu of procedures that could be tested, replicated or refuted. This restriction has contributed to an underlying weakness in the science of impact assessment. One possible solution to this malaise identified by Storey (1986) is that more emphasis needs to be paid to the *management* of project outcomes than to their estimation.

This change requires that impact assessment be placed within a wider frame of reference. In Chapter 6, impact assessment is redefined within an integrated framework for sustainable resource management. The framework is based upon a redefinition of impact assessment as an adaptive, integrative and interactive means of decision making in environmental planning.

As the application of the framework in Chapters 7–9 illustrates, the shift towards a new role for impact assessment requires that several factors need to be addressed. These include:

- impact assessment as a political process of decision making and its defined role within the institutional arrangements for management
- the incorporation of stakeholders into impact assessment processes
- more emphasis upon methods for impact management than impact identification.

Impact assessment as a political process

Impact assessment is not a purely technical exercise. It provides for environmental analysis and is strengthened by adherence to the rules and systematic rigour of science. But many aspects of impact assessment revolve around value choices that are inherently political in nature. Impact assessment involves the governance of resources. Decisions resulting from impact assessments concern the determination and allocation of costs and benefits within society. Ultimately, these choices are political decisions influenced more by cultural attitudes, socio-economic conditions and institutional variables than they are by scientific information.

From Chapter 1 onward, one theme throughout this book is that the political dimensions of impact assessment are just as important to its practice as the rules of science. In redefining impact assessment as a bridge between the science of environmental analysis and the politics of resource management, the framework in Chapter 6 recognizes the pivotal role played by institutional arrangements in the realization of the political dimensions of impact assessment.

The institutional arrangements for resource management directly influence the way in which impact assessment provisions are implemented. These provisions may be modified, amended or curtailed through a range of leverage points, within which the success of impact assessment is influenced by:

- political commitment
- the strength of the legislative base
- the intended role and function of impact assessment
- the roles played by various stakeholders.

At the same time, there is an ongoing need for further research to:

- evaluate existing institutional arrangements for management
- indicate the key components within each leverage point for the effective implementation of impact assessment.

In Chapter 7, the absence of political commitment and adequate institutional arrangements are recognized as major barriers to the application of any impact assessment procedures. Typically, these constraints reflect a lack of consensus concerning the benefits of impact assessment and the role it should play in planning resource developments. In contrast, Chapter 8 illustrates that enabling legislation can facilitate an effective impact assessment process but that successful implementation is still reliant upon the actions and reactions of the proponent, management agencies and affected parties.

This limitation suggests that while the institutional arrangements for management are an important determinant of impact assessment, their effective design may be an elusive goal. One solution to this dilemma is the use of formal guidelines to outline the impact assessment process. These guidelines can be used to give directions to proponents in differing resource sectors as to the expectations for planning through impact assessment. Chapter 9 indicates that such guidelines for impact assessment should incorporate a more overtly political approach to planning. Hence, impact assessment must:

- make explicit provisions for stakeholder involvement
- pay more attention to the *process* of planning
- focus on the derivation of a broadly acceptable *strategy for impact management* rather than the production of a technically precise impact statement.

In addition, guidelines need to specify the consideration of cumulative effects and require the establishment of a monitoring programme for effects, compliance and public concerns from the scoping stage through to plan implementation and facility operation.

Incorporating stakeholders into impact assessment

There is growing recognition that impact assessment must account for the stakeholders in any issue, accommodate their goals and utilize approaches that empower communities within decision making. Moreover, stakeholders must be given a role in defining not just the substantive content of planning but the planning approach itself.

These changes constitute a more overtly political thrust for impact assessment, wherein the process of stakeholder involvement is as important as the substantive nature of the effects being assessed. This shift in emphasis is consistent with the findings of the case study chapters that:

- There needs to be formal and explicit requirement for affected stakeholders to have representation within planning and decision making processes (Ch. 7).

- The reliance upon professional judgement in impact assessments needs to be supplanted through the use of consensual methods of decision making and the use of integrated approaches to interest representation (Ch. 8).
- Impact assessments could be improved through the adoption of provisions requiring that stakeholders be involved at the scoping phase of an impact assessment to determine the nature of the problem to be solved and the planning approach to be utilized. Interest representation should then be facilitated throughout the planning process, with co-ordination ensuring the incorporation of social acceptability criteria with criteria for environmental acceptability (Ch. 9).

In each of these instances, further research is required to determine how these goals for interest representation can be realized. At the same time, researchers must account for the outstanding issues that constrain interest representation by addressing power relationships between stakeholders, concerns over equity, justice and determination of the public interest, and the need for improved skills and training if interest representation is to foster significant contributions by stakeholders to impact assessment.

Developing methods for impact management

The three case study chapters highlight the need for improvements in impact assessment methods. Some substantive changes include the need to:

- develop more appropriate methods for scoping, prediction and significance assessment (Ch. 7)
- improve the scientific quality of impact assessments by explicitly distinguishing as separate activities the prediction of impacts, the measurement of their significance, and the evaluation of their effects (Ch. 8)
- de-emphasize the reliance upon compensation in project mitigation (Ch. 8)
- revise the approaches used in scoping and in the weighting of criteria at the evaluation stage (Ch. 9)
- revise the requirements for, and approaches to, the monitoring of effects, compliance and public concerns (Ch. 9).

More fundamentally, these changes highlight the need for more emphasis upon methods for *impact management* than on those for *impact identification* within impact assessment. As the review in Chapter 2 indicates, existing methods for impact assessment have evolved from a perspective of impact identification. As a group, they tend to be insufficient for the tasks of prediction, significance assessment and evaluation, and they have often proven inadequate for the purposes of interest representation within impact assessment. Thus, while the basic science of impact assessment has often been criticized for its poor quality, little has been done to refocus the basic methods for, and methodology of, impact assessment away from the identification of effects and towards the management of impacts.

At the same time, the preoccupation with impact identification has precluded attention to the need for improved methods for monitoring and mitigating impacts when they do occur. Present methods are rudimentary and there is a shortage of studies that have evaluated experiences with mitigation and monitoring as impact management strategies (e.g. Storey et al, 1991). Thus, the need exists for further research to:

- consider the potential of such existing methods for impact assessment as geographic information systems (GIS) to provide for improved monitoring and mitigation of impacts
- review the utility of existing approaches to monitoring and mitigation as strategies for impact management
- identify and develop new methods for impact assessment that emphasize the management of impacts.

Prospects for change

A wide range of experiences are reviewed in Chapters 7–9. The examples encompass experiences in both developed and developing nations from around the world and cover various levels of planning for an extensive array of topics. The reforms suggested above reflect the breadth of this experience and are intended as generic changes to the manner in which impact assessment is conceived and implemented. Exactly how these changes can be realized in practice will depend upon the specific context of each situation and could vary enormously from nation to nation.

For example, in countries such as the United States or Canada, where the existing paradigm for impact assessment is well entrenched and there is often considerable inertia to institutional reform, the changes advocated in this book could face significant barriers. Similarly, many European countries have already had to adjust to the implications of the EC Directive 85/337 and may be reluctant to further recast their impact assessment provisions. Conversely, in countries where impact assessment is still evolving (such as India) or where revisions have already been based on the concept of sustainability (such as New Zealand), the transition to a new conceptualization of impact assessment should be easier.

This caveat aside, the changes advocated here are considered to be broadly applicable to both developed and developing nations. Similarly, they apply equally to proposed projects, to efforts to rehabilitate degraded habitats and to the remediation of existing problems. The basic tenet is that impact assessment needs to be redefined. The principles for change suggested in this book would result in impact assessments that:

- foster sustainable resource management
- provide a means for environmental planning

Implications for sustainable resource management

- involve stakeholders in the development of a strategy for impact management.

Wherever impact assessment is implemented, it would be improved by these changes.

Bibliography

Abracosa, R. and L. Ortolano (1987) Environmental impact assessment in the Philippines: 1977–1985. *EIA Review*, **7** (3), 293–310.

Amy, D. J. (1990) Environmental dispute resolution: the promise and the pitfalls, in N. J. Vig and M. E. Kraft (eds) *Environmental Policy in the 1990s*. CQ Press: Washington, DC, 211–34.

Anderson, F.R. (1973) *NEPA in the Courts*. Johns Hopkins University Press: Baltimore.

Andrews, R. N. L. (1990) Risk assessment: regulation and beyond, in N. J. Vig and M. E. Kraft (eds) *Environmental Policy in the 1990s*. CQ Press: Washington, DC, 167–86.

Archibugi, F. and P. Nijkamp (eds) (1989) *Economy and Ecology: Towards Sustainable Development*. Kluwer Academic: Dordrecht.

Aronoff, S. (1989) *Geographic Information Systems: A Management Perspective*. WDL Publications: Ottawa.

Bacow, L. S. and M. Wheeler (1984) *Environmental Dispute Resolution*. Plenum: New York.

Bailey, J. and V. Hobbs (1990) A proposed framework and database for EIA auditing. *Journal of Environmental Management*, **31**, 163–72.

Bailey, R.G. (1988) Problems with using overlay mapping for planning and their implications for geographic information systems. *Environmental Management*, **12** (1), 11–17.

Baldwin, J. H. (1985) *Environmental Planning and Management*. Westview: Boulder, Co.

Barrett, S. and M. Hill (1984) Policy, bargaining and structure in implementation theory. *Policy and Politics*, **12** (3), 219–40.

Bartlett, R. V. (ed) (1988) Symposium: policy and impact assessment. *Impact Assessment Bulletin*, **6** (3/4).

Bartlett, R. V. (ed) (1989a) *Policy Through Impact Assessment*. Greenwood: Westport, Conn.

Bartlett, R. V. (1989b) Impact assessment as a policy strategy, in R. V. Bartlett (ed) *Policy Through Impact Assessment*. Greenwood: Westport, Conn., 1–4.

Bartlett, R. V. and W. F. Baber (1989) Bureaucracy or analysis: implications of impact assessment for public administration, in R. V. Bartlett (ed) *Policy Through Impact Assessment*. Greenwood: Westport, Conn., 143–53.

Basta, D. and B. Bower (eds) (1982) *Analyzing Natural Systems*. Resources for the Future: Washington, DC.

Bibliography

Beanlands, G. E. and P. N. Duinker (1983) *An Ecological Framework for Environmental Impact Assessment in Canada*. Institute for Resource and Environmental Studies, Dalhousie University: Halifax.

Beanlands, G. E. and P. N. Duinker (1984) An ecological framework for environmental impact assessment. *Journal of Environmental Management*, **18**, 267–77.

Bebler, A. and J. Seroka (eds) (1990) *Contemporary Political Systems: Classifications and Typologies*. Lynne Rienner: Boulder, Co.

Bendix, J. and C. M. Liebler (1991) Environmental degradation in Brazilian Amazonia: perspectives in US news media. *The Professional Geographer*, **43** (4), 474–85.

Benveniste, G. (1981) *Regulation and Planning: the Case of Environmental Politics*. Boyd and Fraser: San Francisco.

Benveniste, G. (1989) *Mastering the Politics of Planning*. Jossey-Bass: San Francisco.

Berger, T. R. (1976) The Mackenzie Valley pipeline inquiry. *Queen's Quarterly*, **83**, 1–12.

Berger, T. R. (1977) *Northern Frontier, Northern Homeland: The Report of the Mackenzie Valley Pipeline Inquiry* (2 vols). Supply and Services Canada: Ottawa.

Bergstrom, L. (1970) What is a conflict of interest? *Journal of Peace Research*, **7**, 199–217.

Berkes, F. (1988) The intrinsic difficulty of predicting impacts: lessons from the James Bay Hydro project. *EIA Review*, **8**, 201–20.

Berry, J. K. (1987) Computer-assisted map analysis: potential and pitfalls. *Photogrammetric Engineering and Remote Sensing*, **53** (10), 1405–10.

Bidol, P. and M. Lesnick (1984) Environmental conflict management. *Focus*, **9** (1/2).

Bingham, G.(1986) *Resolving Environmental Disputes: A Decade of Experience*. The Conservation Foundation: Washington, DC.

Bisset, R. (1983a) Introduction to methods for environmental impact assessment, in PADC, EIA and Planning Unit (ed) *Environmental Impact Assessment*. Martinus Nijhoff: The Hague, 131–47.

Bisset, R. (1983b) A critical survey of methods for environmental impact assessment, in T. O'Riordan and R. K. Turner (eds) *An Annotated Reader in Environmental Planning and Management*. Pergamon: Oxford, 168–86.

Bisset, R. (1987) Methods for environmental impact assessment: a selective survey with case studies, in A. K. Biswas and Q. Geping (eds) *Environmental Impact Assessment for Developing Countries*. Tycooly International: London, 5–64.

Bisset, R. (1988) Developments in EIA methods, in P.Wathern (ed) *Environmental Impact Assessment: Theory and Practice*. Unwin Hyman: London, 47–61.

Bissett, R. and P. Tomlinson (1983) Environmental impact assessment, monitoring and post-development audits, in PADC, EIA and Planning Unit ed. *Environmental Impact Assessment*. Martius Nijhoff: The Hague, 405–26.

Biswas, A. K. and Q. Geping (eds) (1987) *Environmental Impact Assessment for Developing Countries*. Tycooly International: London.

Biswas, A. K. et al (eds) (1990) *Environmental Modelling for Developing Countries*. Tycooly International: London.

Blacksell, M. and A. Gilg (1981) *The Countryside: Planning and Change*. George Allen and Unwin: London.

Bolan, R. S. (1983) The promise and perils of a normative theory of planning, in I. Masser (ed) *Evaluating Urban Planning Efforts*. Gower: Aldershot, Hampshire, 3–15.

Bonoma, T. (1976) Conflict, co-operation and trust. *Behavioral Science*, **21**, 499–514.

Boothroyd, P. and W. E. Rees (1984) *Impact Assessment: From Pseudo-Science to Planning Process*. School of Community and Regional Planning, University of British Columbia, UBC Planning Papers, DP No. 3: Vancouver, BC.

Bowman, A. O'M. (1988) Superfund implementation, in C. E. Davis and J. P. Lester (eds) *Dimensions of Hazardous Waste Politics and Policy*. Greenwood Press: Westport, Conn., 129–46.

Boyce, J. K. (1990) Birth of a megaproject: political economy of flood control in Bangladesh. *Environmental Management*, **14** (4), 419–28.

Brammer, H. (1990) Floods in Bangladesh: II. Flood mitigation and environmental aspects. *The Geographical Journal*, **156** (2), 158–65.

Braybrooke, D. and C. E. Lindblom (1963) *A Strategy of Decision*. The Free Press, New York.

Brown, A. L. et al (1991) Environmental assessment procedures and issues in the Pacific Basin – Southeast Asia region. *EIA Review*, **11** (2), 143–56.

Brown, B. J. et al (1987) Global sustainability: toward definition. *Environmental Management*, **11** (6), 713–19.

Burrough, P. A. (1986) *Principles of Geographical Information Systems for Land Resource Assessment*. Clarendon Press: Oxford.

Caldwell, L. K. (1982) *Science and the National Environmental Policy Act: Redirecting Policy Through Procedural Reform*. University of Alabama Press: Alabama.

Caldwell, L. K. (1988a) Environmental impact analysis (EIA): origins, evolution, and future directions. *Impact Assessment Bulletin*, **6**, (3–4), 75–83.

Caldwell, L. K. (ed) (1988b) *Perspectives on Ecosystem Management for the Great Lakes*. State University of New York Press: Albany, NY.

Caldwell, L. K. (1989) Understanding impact analysis: technical process, administrative reform, policy principle, in R.V. Bartlett (ed) *Policy Through Impact Assessment*. Greenwood: Westport, Conn., 7–16.

Canales, L. D. (1989) First environmental impact assessment of a highway in Portugal. *EIA Review*, **9**, 391–97.

Canter, L. W. (1983) Methods for environmental impact assessment: theory and application, in PADC, EIA and Planning Unit (ed) *Environmental Impact Assessment*. Martinus Nijhoff: The Hague, 165–227.

CARC (Canadian Arctic Resources Committee) (1984) Not with a bang but with a BEARP. *Northern Perspectives*, **12** (3), 1–3.

Carley, M. J. and E. S. Bustelo (1984) *Social Impact Assessment and Monitoring*. Westview: Boulder, Co.

Carlisle, T. J. and L. G. Smith (1989) Examining the potential for a negotiated approach to water management in Ontario, *Canadian Water Resources Journal*, **14** (4), 16–34.

Cayer, N. J. and L. F. Weschler (1988) *Public Administration: Social Change and Adaptive Management*. St Martin's Press: New York.

Checkoway, B. (1981) The politics of public hearings. *Journal of Behavioral Science*, **17**, 566–82.

Checkoway, B. (ed) (1986) *Strategic Perspectives on Planning Practice*. D. C. Heath: Lexington, Mass.

Cheremisinoff, P. W. and A. C. Morres (1977) *Environmental Assessment and Impact Statement Handbook*. Ann Arbor Science Publishers: Ann Arbor, Mich.

Clark, B. D. (1983a) The aims and objectives of environmental impact assessment, in PADC, EIA and Planning Unit (ed) *Environmental Impact Assessment*. Martinus Nijhoff: The Hague, 3-11.

Clark, B. D. (1983b) EIA manuals: general objectives and the PADC manual, in PADC, EIA and Planning Unit (ed) *Environmental Impact Assessment*. Martinus Nijhoff: The Hague, 149–64.

Clark, B. D. et al (1981) *A Manual for the Assessment of Major Development Proposals*. HMSO: London.

Clark, B. D. et al (eds) (1984) *Perspectives on Environmental Impact Assessment*. D.Reidel: Dordrecht.

Bibliography

Clark, M. and J. Herington (eds) (1988)*The Role of Environmental Impact Assessment in the Planning Process*. Mansell: London.

Clark, W. C. (1986) Sustainable development of the biosphere: themes for a research program, in W. C. Clark and R. E. Munn (eds) *Sustainable Development of the Biosphere*. Cambridge University Press: Cambridge, 5–48.

Clark, W. C. and R. E.Munn (eds) (1986) *Sustainable Development of the Biosphere*. Cambridge University Press: Cambridge.

Coates, V. T. and J. F. Coates (1989) Making technology assessment an effective tool to influence policy, in R. V. Bartlett (ed) *Policy Through Impact Assessment*. Greenwood: Westport, Conn., 17–25.

Colborn, T. E. et al (1990) *Great Lakes, Great Legacy*? The Conservation Foundation and the Institute for Research on Public Policy: Washington, DC and Ottawa.

Coleman, W. D. (1985) Analyzing the associative action of business: policy advocacy and policy participation. *Canadian Public Administration*, **28** (3), 413–33.

Concannon, K. A. (1984) A public involvement strategy for siting transmission lines, in Environment Canada *Facility Siting and Routing '84: Energy and Environment*, 2 vols. Supply and Services Canada: Ottawa, 645–55.

Cope, D. and P. Hills (1988) Total assessment: myth or reality? in M. Clark and J. Herington (eds) *The Role of Environmental Impact Assessment in the Planning Process*. Mansell: London, 174–93.

Cormick,, G. W. (1980) The 'theory' and practice of environmental mediation. *Environmental Professional*, **2**, 24–33.

Craine, L. E. (1971) Institutions for managing lakes and bays. *Natural Resources Journal*, **11**, 519–46.

Curran, P. J. (1987) Remote sensing methodologies and geography. *International Journal of Remote Sensing*, **8** (9), 1255–75.

Daly, H. E. (1990) Toward some operational principles of sustainable development. *Ecological Economics*, **2**, 1–6.

Darke, R. (1983) Procedural planning theory, in I. Masser (ed) *Evaluating Urban Planning Efforts*. Gower: Aldershot, Hampshire, 16–35.

Davis, C. E. and J. P. Lester (1988) Hazardous waste policies and the policy process, in C. E. Davis and J. P. Lester (eds) *Dimensions of Hazardous Waste Politics and Policy*. Greenwood Press: Westport, Conn., 1–34.

de Broissia, M. (1986) *Selected Mathematical Models in Environmental Impact Assessment in Canada*. Supply and Services Canada: Ottawa.

Dee, N. et al (1973) An environmental evaluation system for water resource planning. *Water Resources Research*, 9, 523–35.

Dienemann, E. A. et al (1991) Remediation of the Lipari Landfill, America's #1 ranked Superfund site. *Impact Assessment Bulletin*, **9** (3), 13–30.

Doern, G. B. and R. W. Phidd (1983) *Canadian Public Policy: Ideas, Structure, Process*. Methuen: Toronto.

Dorcey, A.H.J. (1986) *Bargaining in the Governance of Pacific Coastal Resources*. Westwater Research Centre, University of British Columbia: Vancouver.

Dorney, R. S. (1977) Environmental assessment: the ecological dimension. *Journal of the American Waterworks Association*, **69**, 182–5.

Dovers, S. R. (1990) Sustainability in context: an Australian perspective. *Environmental Management*, **14** (3), 297–305.

Dower, R. C. (1990) Hazardous wastes, in P. K. Portney et al (eds) *Public Policies for Environmental Protection*. Resources for the Future: Washington, DC, 151–94.

Downs, A. (1972) Up and down with ecology. *The Public Interest*, **28**, 38–50.

Ducsik, D. W. (1984) Power plants and people: a profile of electric utility initiatives in cooperative planning. *Journal of the American Planning Association*, **50** (2), 162–74.

194

Easton, D. (1965) *A Systems Analysis of Political Life*. John Wiley: New York.

Edwards, C. J. and H. A. Regier (eds) (1990) *An Ecosystem Approach to the Integrity of the Great Lakes in Turbulent Times*. Great Lakes Fishery Commission Special Publication 90–4: Ann Arbor, Mich.

ESSA (Environmental and Social Systems Analysts) Ltd (1982) *Review and Evaluation of Adaptive Environmental Assessment and Management*. Environment Canada: Ottawa.

Estes, J. E. et al (1987) Coordinating hazardous waste management activities using geographical information systems. *International Journal of Geographical Information Systems*, **1**, (4), 359–77.

Everest, D. A. (1990) The provision of expert advice to government on environmental matters. *Science and Public Affairs*, **4**, 17–40.

Everitt, R. R. (1983) Adaptive environmental assessment and management: some current applications, in PADC, EIA and Planning Unit (ed) *Environmental Impact Assessment*. Martinus Nijhoff: The Hague, 293–306.

Fagence, M. (1977) *Citizen Participation in Planning*. Pergamon: Oxford.

Fairfax, S. K. (1978) A disaster in the environmental movement. *Science*, **199**, 743.

Fairfax, S. K. and H. Ingram (1981) The United States experience, in T. O'Riordan and W. R. D. Sewell (eds) *Project Appraisal and Policy Review*. John Wiley: Chichester, 29–45.

Faludi, A. (1973) *Planning Theory*. Pergamon: Oxford.

Faludi, A. (1987) *A Decision-centred View of Environmental Planning*. Pergamon: Oxford.

Fearnside, P. M. (1990) Environmental destruction in the Brazilian Amazon, in D. Goodman and A. Hall (eds) *The Future of Amazonia*. Macmillan: London, 179–225.

Fenge, T. (1984) Measuring up: did the BEARP fulfil expectations? *Northern Perspectives*, **12** (3), 14–16.

Fenge, T. and L. G. Smith (1986) Reforming the Federal Environmental Assessment and Review Process. *Canadian Public Policy*, **12** (4), 596–605.

Finsterbusch, K. (1980) *Understanding Social Impacts*. Sage: Beverly Hills.

Finsterbusch, K. (1989) Community responses to exposure to hazardous wastes, in D. L. Peck (ed) *Psychosocial Effects of Hazardous Toxic Waste Disposal on Communities*. Charles C. Thomas: Springfield, Ill., 57–80.

Finsterbusch, K. et al (eds) (1983) *Social Impact Assessment Methods*. Sage: Beverly Hills.

Finsterbusch, K. and C. P. Wolf (eds) (1977) *Methodology of Social Impact Assessment*. Dowden, Hutchinson and Ross: Stroudsburg, Pa.

Fisher, R. and W. Ury (1981) *Getting to Yes*: *Negotiating Without Giving In*. Houghton Mifflin: Boston.

Forester, W. S. and J. H. Skinner (eds) (1987) *International Perspectives on Hazardous Waste Management*. Academic Press: London.

Formby, J. (1981) The Australian experience, in T. O'Riordan and W. R. D. Sewell (eds) *Project Appraisal and Policy Review*. John Wiley: Chichester, 187–226.

Fowle, C. D. et al (eds) (1988) *Information Needs for Risk Management*. Institute for Environmental Studies, University of Toronto: Toronto.

Frank, A. G. (1966) The development of underdevelopment. *Monthly Review*, **18** (4), 17–31.

Freeman, A. M. (1990) Water pollution policy, in P.K. Portney et al (eds) *Public Policies for Environmental Protection*. Resources for the Future: Washington, DC, 97–149.

Friedmann, J. (1973) *Retracking America*: *A Theory of Transactive Planning*. Doubleday: New York.

Friere, P. (1970) *Pedagogy of the Oppressed*. Seabury: New York.

Friesma, P and P. Culhane (1976) Social impacts, politics, and the environmental impact statement process. *Natural Resources Journal*, **16** (2), 339–56.

Gardner, J. E.(1988) Decision-making for sustainable development, in J. E. Gardner, W. E. Rees and P.Boothroyd *The Role of Environmental Assessment in Promoting Sustainable Development*. School of Community and Regional Planning, University of British Columbia, UBC Planning Papers, DP No. 13: Vancouver, BC, 1–19.

Gawthorp, L. C. (1983) Organizing for change. Annals. *American Academy of Political and Social Science*, 466, 119–34.

Gibbs, L. M. (1982) *Love Canal: My Story*. State University of New York Press: Albany, NY.

Gibson, R. B. (1990) Lessons of a legislated process. *Impact Assessment Bulletin*, **8** (3), 63–80.

Gibson, R. B. and B. Savan (1986) *Environmental Assessment in Ontario*. Canadian Environmental Law Research Foundation: Toronto.

Given, G. L. (1989) Monitoring Public Concerns in Impact Assessment, MA Thesis, Dept. of Geography, University of Western Ontario: London, Ont.

Goldemberg, J. and E. R. Durham (1990) Amazônia and national sovereignty. *International Environmental Affairs*, **2** (1), 22–39.

Goldsmith, E. and N. Hildyard (1984) *The Social and Environmental Effects of Large Dams*. Sierra Club Books: San Francisco.

Goodland, R. J. A. (1990a) Environment and development: progress of the World Bank. *The Geographical Journal*, **156** (2), 149–57.

Goodland, R. J. A. (1990b) The World Bank's new environmental policy on dam and reservoir projects. *International Environmental Affairs*, **2** (2), 109–27.

Goodman, D. and A. Hall (eds) (1990a) *The Future of Amazonia*. Macmillan: London.

Goodman, D. and A. Hall (1990b) Introduction, in D. Goodman and A. Hall (eds) *The Future of Amazonia*. Macmillan: London, 1–20.

Gormley, W. T. (1987) Institutional policy analysis: a critical review. *Journal of Policy Analysis and Management*, **6** (2), 153–69.

Gormley, W. T. (1989) *Taming the Bureaucracy*. Princeton University Press: Princeton, NJ.

Grad, F. P. (1985) *Environmental Law*, 3rd edn. Mathew Bender: New York.

Grima, A. P. et al (1986) *Risk Management and EIA: Research Needs and Opportunities*. Supply and Services Canada: Ottawa.

Grima, A. P. et al (eds) (1989) *Risk Perspectives on Environmental Impact Assessment*. Institute for Environmental Studies, University of Toronto: Toronto.

Grima, A. P. and C. D. Fowle (1989) Reflections and conclusions, in A. P. Grima et al (eds) *Risk Perspectives on Environmental Impact Assessment*. Institute for Environmental Studies, University of Toronto: Toronto, 151–60.

Hague, R. and M. Harrop (1987) *Comparative Government and Policies: An Introduction*, 2nd edn. Macmillan: London.

Hall, A. L. (1989) *Developing Amazonia*. Manchester University Press: Manchester.

Hallman, W. and A. Wandersman (1989) Perception of risk and toxic hazards, in D. L. Peck (ed) *Psychosocial Effects of Hazardous Toxic Waste Disposal on Communities*. Charles C. Thomas: Springfield, Ill., 31–56.

Hecht, S. and A. Cockburn (1990) *The Fate of the Forest*. Penguin: London.

Hertig, J-A. (1990) Méthodologie de l'étude d'impact sur l'environnement d'un projet routier. *IAIA'90, Proceedings of the International Association for Impact Assessment*, Lausanne, Switzerland, 90–3.

Hill, M. A. (1968) A goals–achievement matrix for evaluating alternative plans. *Journal of the American Institute of Planners*, **34**, 19–28.

Hirst, S. M. (1984) Applied ecology and the real world. *Journal of Environmental Management*, **18**, 189–213.

Holling, C. S. (ed) (1978) *Adaptive Environmental Assessment and Management*. John Wiley: Chichester.

Horberry, J. (1985) International organization and EIA in developing countries. *EIA Review*, **5** (3), 207–22.

Hudson, B. M. (1979) Comparison of current planning theories: counterparts and contradictions. *Journal of the American Planning Association*, **45**, 387–98.

Hyman, E. L. and B. Stiftel (1988) *Combining Facts and Values in Environmental Impact Assessment*. Westview Press: Boulder, Co.

Independent Commission on International Development Issues (1980) *North–South: A Programme for Survival*. Pan: London.

Ingram, H. M. et al (1984) Guidelines for improved institutional analysis in water resources planning. *Water Resources Research*, **20**, 323–34.

International Union for the Conservation of Nature and Natural Resources (1980) *World Conservation Strategy*. IUCN: Gland, Switzerland.

Jacobs, H. M. and R. Rubino (1987) Environmental mediation: an annotated bibliography. *CPL Bibliography*, No. 189, 1–24.

Jacobs, P. and D. A. Munro (eds) (1987) *Conservation with Equity: Strategies for Sustainable Development*. Cambridge University Press: Cambridge.

Jain, R. K. et al (1977) *Environmental Impact Analysis*. Van Nostrand Reinhold: New York.

Jakimchuk, R.D. (1987) Follow-up to environmental assessment for pipeline projects in Canada, in B. Sadler (ed) *Audit and Evaluation in Environmental Assessment and Management*, vol.1. Supply and Services Canada: Ottawa, 32–45.

Jenkins, W. I. (1978) *Policy Analysis: A Political and Organizational Perspective*. Martin Robertson: London.

Jéquier, N. and C. Weiss (1984a) Introduction: the World Bank as a technological institution, in C. Weiss and N. Jéquier (eds) *Technology, Finance, and Development: An Analysis of the World Bank as a Technological Institution*. Lexington Books: Lexington, Mass., 1–13.

Jéquier, N. and C. Weiss (1984b) General conclusions, in C. Weiss and N. Jéquier (eds) *Technology, Finance, and Development: An Analysis of the World Bank as a Technological Institution*. Lexington Books: Lexington, Mass., 313–28.

Johnston, C. A. et al (1988) Geographic information systems for cumulative impact assessment. *Photogrammetric Engineering and Remote Sensing*, **54** (11), 1609–15.

Johnston, R. A. and W. S. McCartney (1991) Local government implementation of mitigation requirements under the California Environmental Quality Act. *EIA Review*, **11** (1), 53–67.

Jones, C. A. (1983) *The North–South Dialogue: A Brief History*. Francis Printer: London.

Jones, M. L. and L. A. Greig (1985) Adaptive environmental assessment and management: a new approach to environmental impact assessment, in V. W. Maclaren and J. B. R. Whitney (eds) *New Directions in Environmental Impact Assessment in Canada*. Methuen: Toronto, 21–42.

Kasperson, R. E. (1969) Political behaviour and the decision-making process in the allocation of water resources between recreational and municipal uses. *Natural Resources Journal*, 9, 176–211.

Kates, R. W. and I. Burton (eds) (1986) *Geography, Resources, and Environment*: vol.1, *The Selected Writings of Gilbert F. White*. Chicago University Press: Chicago.

Keeney, R. L. (1980) *Siting Energy Facilities*. Academic Press: New York.

Keeney, R. L. and H. Raiffa (1976) *Decisions With Multiple Objectives*. Wiley: New York.

Kemp, R., T. O'Riordan and M. Purdue (1984) Investigation as legitimacy: the maturing of the big public inquiry. *Geoforum*, **15** (3), 477–88.

Kennedy, W. V. (1988a) Environmental impact assessment in North America, Western Europe: what has worked, where, how and why. *International Environmental Reporter*, **11** (4), 257–62.

Kennedy, W. V. (1988b) Environmental impact assessment and bilateral development aid: an overview, in P.Wathern (ed) *Environmental Impact Assessment: Theory and Practice*. Unwin Hyman: London, 272–85.

Kidron, M. and R. Segal (1984) *The New State of the World Atlas*. Pan: London.

King, J. E. and J. G. Nelson (1983) Evaluating the Federal Environmental Assessment and Review Process with reference to South Davis Strait, Northeastern Canada. *Environmental Conservation*, **10** (4), 293–301.

Kraft, M. E. (1988) Analyzing technological risks in Federal regulatory agencies, in M. E. Kraft and N. J. Vig (eds) *Technology and Politics*. Duke University Press: Durham, NC, 184–207.

Krawetz, N. M. et al (1987) *A Framework for Effective Monitoring*. Supply and Services Canada: Ottawa.

Lake, R. W. (ed) (1987) *Resolving Locational Conflict*. Centre for Urban Policy Research, Rutgers University: New Brunswick, N.J.

Lang, R. (ed) (1986a) *Integrated Approaches to Resource Planning and Management*. University of Calgary Press: Calgary.

Lang, R. (1986b) Achieving integration in resource planning, in R. Lang (ed) *Integrated Approaches to Resource Planning and Management*. University of Calgary Press: Calgary, 27–50.

Langton, S. (ed) (1978) *Citizen Participation in America*. Lexington: Toronto.

Langton, S. (ed) (1979) *Citizen Participation Perspectives*. Lincoln Filene Centre for Citizenship and Public Affairs, Tufts University: Medford, Mass.

Lasswell, H. D. (1950) *Politics: Who Gets What, When, How*. Peter Smith: New York.

Lee, W. and C. Wood (1978) Environmental impact assessment of projects in EEC countries. *Journal of Environmental Management*, **6** (1), 57–71.

Leistritz, F. L. and S. H. Murdock (1981) *The Socioeconomic Impact of Resource Development: Methods for Assessment*. Westview Press: Boulder, Co.

Le Marquand, D. (1989) Developing river and lake basins for sustained economic growth and social progress. *Natural Resources Forum*, **13** (2), 127–38.

Leopold, L. et al (1971) *A Procedure for Evaluating Environmental Impact*. Survey Circular 645, US Geological Survey: Washington, DC.

Le Prestre, P. (1989) *The World Bank and the Environmental Challenge*. Associated University Press: London.

Lester, J. P. (1988) Superfund implementation: exploring environmental gridlock. *EIA Review*, **8** (2), 159–74.

Levine, A. (1982) *Love Canal: Science, Politics and People*. D. C. Heath: Lexington, Mass..

Lijphart, A. (1990) Democratic political systems, in A. Bebler and J. Seroka (eds) *Contemporary Political Systems: Classifications and Typologies*. Lynne Rienner: Boulder, Co., 71–87.

Lindblom, C. E. (1959) The science of muddling through. *Public Administration Review*, **19**, 79–88.

Lindblom, C. E. (1979) Still muddling, not yet through. *Public Administration Review*, **39**, 517–26.

Lohani, B. N. and N. Halim (1987) Recommended methodologies for rapid environmental impact assessment in developing countries, in A. K. Birwas and Q. Geping (eds) *Environmental Impact Assessment for Developing Countries*. Tycooly International: London, 65–111.

Lowe, P. and J. Goyder (1983) *Environmental Groups in Politics*. George Allen and Unwin: London.

Maass, A. (1951) *Muddy Waters: The Army Engineers and the Nation's Rivers*. Harvard University Press: Cambridge, Mass.

McAllister, D.M. (1980) *Evaluation in Environmental Planning*. MIT Press: Cambridge, Mass..

McAuslan, P. (1980) *The Ideologies of Planning Law*. Oxford: Pergamon.

McHarg, I. (1969) *Design with Nature*. Doubleday: Garden City, NY.

McKechnie, R. et al (1983) Sanitary landfills and their impact upon land, in W. Simpson-Lewis et al (eds) *Stress on Land in Canada*. Environment Canada: Ottawa.

McKinney, M.J. (1988) Water resources planning: a collaborative, consensus-building approach, *Society and Natural Resources*, **1** (4), 335–50.

Maclaren, V. W. (1991) Waste management: current crisis and future challenge, in B. Mitchell (ed) *Resource Management and Development*. Oxford University Press: Oxford, 28–53.

McQuaid-Cook, J. and K. J. Simpson (1986) Siting a fully integrated waste management facility. *APCA Journal*, **36** (9), 1031–6.

March, J. G. and H. A. Simon (1958) *Organizations*. Wiley: New York.

Marks, J. and L. Susskind (1988) Negotiating better Superfund settlements. *EIA Review*, **8** (2), 113–32.

Marquès, I. (1990) Carajas, in O. Bomsel et al (eds) *Mining and Metallurgy Investment in the Third World: the End of Large Projects?* OECD: Paris, 53–70.

Marshall, D. et al (1985) *Environmental Management and Impact Assessment*. Occasional Paper, Federal Environmental Assessment Review Office, Environment Canada: Ottawa.

Massa, A. K. (1984) Energy facility siting and routing, in Environment Canada *Facility Siting and Routing '84: Energy and Environment*, 2 vols. Supply and Services Canada: Ottawa, 13–23.

Mazmanian, D. and D. Morell (1990) The 'NIMBY' syndrome: facility siting and the failure of democratic discourse, in N. J. Vig and M. E. Kraft (eds) *Environmental Policy in the 1990s*. CQ Press: Washington, DC, 125–43.

Mazmanian, D. and J. Nienaber (1979) *Can Organizations Change?* Brookings Institution: Washington, DC.

Mazur, A. (1989) Communicating risk in the mass media, in D.L. Peck (ed) *Psychosocial Effects of Hazardous Toxic Waste Disposal on Communities*. Charles C. Thomas: Springfield, Ill., 119–38.

Milbraith, L. W. (1963) *The Washington Lobbyists*. Rand McNally: Chicago.

Mishan, E. J. (1976) *Elements of Cost–Benefit Analysis*, 2nd edn. George Allen and Unwin: London.

Mitchell, B. (1986) The evolution of integrated resource management, in R.Lang (ed) *Integrated Approaches to Resource Planning and Management*. University of Calgary Press: Calgary, 13–26.

Mitchell, B. (1987) *A Comprehensive-integrated Approach for Water and Land Management*. Centre for Water Policy Research, University of New England, Occasional Paper No. 1: Armidale, NSW.

Mitchell, B. (1989) *Geography and Resource Analysis*, 2nd edn. Longman: London.

Mitchell, B. (1990a) Integrated water management, in B. Mitchell (ed) *Integrated Water Management: International Experiences and Perspectives*. Belhaven Press: London, 1–21.

Mitchell, B. (1990b) Improved flying without new wings, in R. Y. McNeil and J. E. Windsor (eds) *Innovations in River Basin Management*. Canadian Water Resources Association: Cambridge, Ont., 7–16.

Mitchell, B. (ed) (1990c) *Integrated Water Management: International Experiences and Perspectives*. Belhaven Press: London.

Mitchell, B. and J. J. Pigram (1989) Integrated resource management and the

Hunter Valley Conservation Trust, NSW, Australia. *Applied Geography*, **9**, 196–211.

Moffat, I. (1990) The potentialities and problems associated with applying information technology to environmental management. *Journal of Environmental Management*, **30**, 209–20.

Moncrieff, I. et al (1987) An environmental performance audit of selected pipeline projects in Southern Ontario, in B. Sadler (ed) *Audit and Evaluation in Environmental Assessment and Management*, vol.1 Supply and Services Canada: Ottawa, 145–59.

Morgan, R. K. (1988) Reshaping environmental impact assessment in New Zealand. *EIA Review*, **8**, 293–306.

Munn, R. E. (1983) The theory and application of modelling in environmental impact assessment, in PADC, EIA and Planning Unit (ed) *Environmental Impact Assessment*. Martinus Nijhoff: The Hague, 281–92.

Munro, D. A. et al (1986) *Learning From Experience: A State-of-the-art Review and Evaluation of Environmental Impact Assessment Audits*. Supply and Services Canada: Ottawa.

Nelson, J. G. and S. Jessen (1981) *The Scottish and Alaskan Offshore Oil and Gas Experience and the Canadian Beaufort Sea*. Canadian Arctic Resources Committee: Ottawa.

Neto, F. T. (1990) Development planning and mineral mega-projects, in D. Goodman and A. Hall (eds) *The Future of Amazonia*. Macmillan: London, 130–54.

Nishimura, H. (ed) (1989) *How to Conquer Air Pollution: A Japanese Experience*. Elsevier: Amsterdam.

Noble, J. H. et al (eds) (1977) *Groping through the Maze*. The Conservation Foundation: Washington, DC.

Nor, Y. (1991) Problems and perspectives in Malaysia. *EIA Review*, **11** (2), 129–42.

Odum, H. T. (1971) *Environment, Power and Society*. Wiley: New York.

OECD (1989) *Strengthening Environmental Co-operation with Developing Countries*. Organization for Economic Co-operation and Development: Paris.

Ontario Hydro (1990) *Bulk Transmission West of London Environmental Assessment*. Ontario Hydro Report No. 90252: Toronto.

O'Riordan, T. (1976) Policy making and environmental management, in A. E. Utton, W. R. D. Sewell and T. O'Riordan (eds) *Natural Resources for a Democratic Society*. Westview: Boulder, Co., 55–72.

O'Riordan, T. (1981) *Environmentalism*, 2nd edn. Pion: London.

O'Riordan, T. (1988) The politics of sustainability, in R. K. Turner (ed) *Sustainable Environmental Management*. Belhaven Press: London, 29–50.

O'Riordan, T. (1990) On the 'greening' of major projects. *The Geographical Journal*, **156** (2), 141–8.

O'Riordan, T., R. Kemp and M. Purdue (1988) *Sizewell B: An Anatomy of the Inquiry*. Macmillan: London.

O'Riordan, T. and W. R. D. Sewell (1981) From project appraisal to policy review, in T. O'Riordan and W. R. D. Sewell (eds) *Project Appraisal and Policy Review*. John Wiley: Chichester, 1–28.

O'Riordan, T. and R. K. Turner (eds) (1983) *An Annotated Reader in Environmental Planning and Management*. Pergamon: Oxford.

Ortolano, L. (1984) *Environmental Planning and Decision Making*. John Wiley: New York.

Ozbekhan, H. (1969) Toward a general theory of planning, in E. Jantsch (ed.) *Technological Planning and Social Futures*. Cassell: London, 47–155.

Ozbekhan, H. (1973) The emerging methodology of planning. *Fields Within Fields*, 10, 63–80.

PADC, EIA and Planning Unit (ed) (1983) *Environmental Impact Assessment.* Martinus Nijhoff: The Hague.

Paschen, H. (ed) (1989) *The Role of Environmental Impact Assessment in the Decisionmaking Process.* Erich Schmidt Verlag: Berlin.

Pearce, D. W. and R. K. Turner (1990) *Economics of Natural Resources and the Environment.* Harvester Wheatsheaf: New York.

Peterson, E. B. et al (1987) *Cumulative Effects Assessment in Canada.* Supply and Services Canada: Ottawa.

Phantumvanit, D. and W. Nandhabiwat (1989) The Nam Choan controversy: an EIA in practice. *EIA Review*, **9** (2), 135–47.

Pirages, D. (1978) *Global Ecopolitics: A New Context for International Relations.* Duxbury Press: N. Scituate, Mass.

Pirages, D. (1989) *Global Technopolitics: The International Politics of Technology and Resources.* Brooks/Cole: Pacific Grove, Calif.

Portney, P. K. (1990) Air pollution policy, in P.K. Portney et al (eds) *Public Policies for Environmental Protection.* Resources for the Future: Washington, DC, 27–96.

Portney, P. K. (1991) *Siting Hazardous Waste Treatment Facilities: The Nimby Syndrome.* Auburn House: Westport, Conn.

Pross, A. P. (1986) *Group Politics and Public Policy.* Oxford University Press: Toronto.

Quarantelli, E. L. (1989) Characteristics of citizen groups which emerge with respect to hazardous waste sites, in D.L. Peck (ed) *Psychosocial Effects of Hazardous Toxic Waste Disposal on Communities.* Charles C. Thomas: Springfield, Ill., 177–96.

Raiffa, H. (1982) *The Art and Science of Negotiation.* Harvard University Press: Cambridge, Mass.

Rau, J. G. and D. C. Wooten (1980) *Environmental Impact Assessment Handbook.* McGraw-Hill: New York.

RCEPP (1980) *Final Report of the Royal Commission on Electric Power Planning in Ontario*, 9 vols. Queen's Printer: Toronto.

Redclift, M. (1987) *Sustainable Development: Exploring the Contradictions.* Methuen: London.

Rees, W. E. (1980) EARP at the crossroads: environmental impact assessment in Canada. *EIA Review*, **1** (4), 355–77.

Rees, W. E. (1984) The process: did the BEARP work? *Northern Perspectives*, **12** (3), 4–6.

Rees, W. E. (1988) A role for environmental assessment in achieving sustainable development. *EIA Review*, **8** (3), 273–91.

Rees, W. E. (1990a) The ecology of sustainable development. *The Ecologist*, **20**, 18–23.

Rees, W. E. (1990b) Atmospheric change: human ecology in disequilibrium. *International Journal of Environmental Studies*, **36**, 103–24.

Roberts, L. E. J. and M. R. Hayns (1989) Limitations on the usefulness of risk assessment. *Risk Analysis*, **9** (4), 483–94.

Roberts, R. D. and T. M. Roberts (eds) (1984) *Planning and Ecology.* Chapman and Hall: London.

Rossini, F. A. and A. L. Porter (1983) Why integrated impact assessment? in F. A. Rossini and A. L. Porter (eds) *Integrated Impact Assessment.* Westview: Boulder, Co., 3–16.

Royston, M. G. (1979) *Pollution Prevention Pays.* Pergamon: Oxford.

Royston, M. G. and Perkowski, J. C. (1975) Determination of the priorities of 'actors' in the framework of environmental management. *Environmental Conservation*, **2** (2), 137–44.

Rubin, D. M. and D. P. Sachs (1973) *Mass Media and the Environment*. Praeger: New York.

Sabatier, P. A. and D. A. Mazmanian (1981) The implementation of public policy: a framework of analysis, in D. A. Mazmanian and P. A. Sabatier (eds) *Effective Policy Implementation*. DC Heath: Lexington, Mass., 3–35.

Sadler, B. (ed) (1978) *Involvement and Environment*, 2 vols. Environment Council of Alberta: Edmonton.

Sadler, B. (ed) (1981) *Public Participation in Environmental Decision Making*. Environment Council of Alberta: Edmonton.

Sadler, B. (1986) Impact assessment in transition: a framework for redeployment, in R. Lang (ed) *Integrated Approaches to Resource Planning and Management*. University of Calgary Press: Calgary, 99–129.

Sadler, B. (1990) *An Evaluation of the Beaufort Sea Environmental Assessment Panel Review*. Supply and Services Canada: Ottawa.

Sandbach, F. (1980) *Environment, Ideology and Policy*. Basil Blackwell: Oxford.

Sandbach, F. (1982) *Principles of Pollution Control*. Longman: London.

Sandman, P. M. et al (1987) *Environmental Risk and the Press*. Transaction Books: New Brunswick, NJ.

Sax, J. (1970) *Defending the Environment*. Alfred Knopf: New York.

Schaller, J. (1990) Geographic information system applications in environmental impact assessment, in H. J. Scholten, and J. C. H. Stillwell (eds) *Geographic Information Systems for Urban and Regional Planning*. Kluwer Academic: Dordrecht, 107–17.

Scudder, T. (1989) The African experience with river basin development. *Natural Resources Forum*, **13** (2), 139–48.

Sebastiani, M. et al (1989) Cumulative impact and sequential geographical analysis as tools for land use planning. *Journal of Environmental Management*, **29**, 237–48.

Sen, A. K. (1984) *Resources, Values and Development*. Harvard University Press: Cambridge, Mass.

Sewell, W. R. D. (1973) Broadening the approach to evaluation in resource management decision making. *Journal of Environmental Management*, **1**, 33–60.

Sewell, W. R. D. (1974) Perceptions, attitudes and public participation in country-side management in Scotland. *Journal of Environmental Management*, **2**, 235–57.

Sewell, W. R. D. (1981) How Canada responded: The Berger Inquiry, in T. O'Riordan and W. R. D. Sewell (eds) *Project Appraisal and Policy Review*. John Wiley: Chichester, 77–94.

Sewell, W. R. D. (1987) The politics of hydro-megaprojects. *Natural Resources Journal*, **27** (2), 497–532.

Sewell, W. R. D. and J. T. Coppock (eds) (1977) *Public Participation in Planning*. John Wiley: London.

Sewell, W. R. D. et al (1965) *A Guide to Benefit–Cost Analysis*. Queen's Printer: Ottawa.

Shafritz, J. M. and A. C. Hyde (eds) (1987) *Classics of Public Administration*, 2nd edn. Dorsey: Chicago.

Shearman, R. (1990) The meaning and ethics of sustainability. *Environmental Management*, **14** (1), 1–8.

Sheate, W. R. and R. M. Taylor (1990) The effect of motorway development on adjacent woodland. *Journal of Environmental Management*, **31**, 261–67.

Shopley, J. B. and R. F. Fuggle (1984) A comprehensive review of current environmental impact assessment methods and techniques. *Journal of Environmental Management*, **18**, 25–47.

Simeon, R. (1976) Studying public policy. *Canadian Journal of Political Science*, **IX** (4), 548–80.

Simon, H. A. (1959) Theories of decision-making in economic and behavioral science. *American Economic Review*, **49**, 253–83.

Simon, H. A. (1976) *Administrative Behaviour*, 3rd edn. New York: The Free Press.

Smith, L. G. (1982a) Mechanisms for public participation at a normative planning level in Canada. *Canadian Public Policy*, **8** (4), 561–72.

Smith, L. G. (1982b) Alternative mechanisms for public participation in environmental policy making. *Environments*, **14** (1), 21–34.

Smith, L. G. (1983) Electric power planning in Ontario. *Canadian Public Administration*, **26** (3), 360–77.

Smith, L. G. (1984a) Public participation in policy making. *Geoforum*, **15** (2), 253–59.

Smith, L. G. (1984b) *Institutional Arrangements for Electric Power Planning in Ontario*. Wilfrid Laurier Research Paper Series No. 8472: Waterloo, Ont.

Smith, L. G. (1987a) The evolution of public participation in Canada. *British Journal of Canadian Studies*, **2** (2), 213–35.

Smith, L. G. (1987b) A performance rating for Canadian environmental impact assessment. *The Operational Geographer*, **13**, 12–14.

Smith, L. G. (1988a) Taming B. C. Hydro: Site C and the implementation of the B.C. *Utilities Commission Act. Environmental Management*, **12** (4), 429–43.

Smith, L. G. (1988b) Institutional considerations in stream management: the need for a negotiated approach, in J. Fitzgibbon and P. Mason (eds) *Managing Ontario Streams*. Canadian Water Resources Association: Cambridge, Ont., 157–64.

Smith, L. G. (1989) Evaluating Canadian impact assessment provisions and practice. Paper presented at the 8th Annual Meeting of the International Association for Impact Assessment, Montreal, 24–28 June, 1989.

Smith, L. G. (1990a) The changing dynamics of interest representation in water resources management, in J.E. Fitzgibbon (ed) *International and Transboundary Water Resources Issues*. American Water Resources Association: Bethesda, Md, 27–34.

Smith, L. G. (1990b) Canada's changing impact assessment provisions. *EIA Review*, **11**, (1), 5–9.

Smith, L. G. (1990c) From condescension to conflict resolution: adjusting to the changing role of the public in impact assessment. Paper presented to the Symposium on Social Aspects of Facility Planning, Institute for Social Impact Assessment, Toronto, 1–3 October, 1990.

Smith, L. G. and L. F. Cattrysse (1987) The changing reality of linear utility planning in Ontario. *Ontario Geography*, **30**, 53–89.

Smith, L. G. and G. L. Given (1989) Impact monitoring and mitigation: the need for improvement. Paper presented at the 8th Annual Meeting of the International Association for Impact Assessment, Montreal, 24–28 June, 1989.

Smith, L. G. et al (1989) An evaluation of Canadian impact assessment practice. *Ontario Geography*, **32**, 1–8.

Smith, P. G. R. and J. B. Theberge (1987) Evaluating natural areas using multiple criteria: theory and practice. *Environmental Management*, **11** (4), 447–60.

Sondheim, M. W. (1978) A comprehensive methodology for assessing environmental impact. *Journal of Environmental Management*, **6**, 27–42.

Sonntag, N. C. et al (1987) *Cumulative Effects Assessment: A Context for Further Research and Development*. Supply and Services Canada: Ottawa.

Spain, D. (1986) Women's role in opposing locally unwanted land uses, in *Not In My Backyard! Community Reaction to Locally Unwanted Land Use*. Institute for Environmental Negotiation, University of Virginia: Charlottesville, Va, 33–40.

Stanbury, W. T. (1978) Lobbying and interest group representation in the legislative process, in W.A.W. Neilson and J.C. McPherson (eds) *The Legislative Process in Canada*. Institute for Research on Public Policy: Montreal, 167–226.

Stanbury, W. T. (1986) *Business–Government Relations in Canada*. Methuen: Toronto.

Starkie, D. (1982) *The Motorway Age*. Pergamon: Oxford.

Stern, A. J. (1991) Using environmental impact assessments for dispute resolution. *EIA Review*, **11**, 81–7.

Storey, K. (1986) From prediction to management: increasing the effectiveness of SIA, in H. A. Becker and A. L. Porter (eds) *Impact Assessment Today*, vol. 2. Jan Van Arkel: Utrecht, 539–51.

Storey, K. et al (1991) *Monitoring for Management*. Institute for Social and Economic Research, Memorial University of Newfoundland: St John's, Newfoundland.

Sullivan, T. J. (1984) *Resolving Development Disputes through Negotiations*. Plenum: New York.

Susskind, L. E. et al (1978) *Resolving Environmental Disputes*. MIT Press: Cambridge, Mass.

Suter, G. W. et al (1987) Treatment of risk in environmental impact assessment. *Environmental Management*, **11** (3), 295–303.

Swartzman, D. et al (eds) (1982) *Cost–Benefit Analysis and Environmental Regulations: Politics, Ethics and Methods*. The Conservation Foundation: Washington, DC.

Taylor, S. (1984) *Making Bureaucracies Think*. Stanford University Press: Stanford.

Tester, F. J. and W. Mykes (eds) (1981) *Social Impact Assessment: Theory, Method and Practice*. Detselig: Calgary.

Thompson, L. S. (1984) Compensation of powerline impacts as a cost-effective alternative to relocation, in Environment Canada *Facility Siting and Routing '84: Energy and Environment*, 2 vols. Supply and Services Canada: Ottawa, 688–706.

Thompson, M. A. (1990) Determining impact significance in EIA: a review of 24 methodologies. *Journal of Environmental Management*, **30**, 235–50.

Turner, R. K. (ed) (1988a) *Sustainable Environmental Management: Principles and Practice*. Belhaven Press: London.

Turner, R. K. (1988b) Sustainability, resource conservation and pollution control: an overview, in R. K. Turner (ed) *Sustainable Environmental Management: Principles and Practice*. Belhaven Press: London, 1–28.

Tyme, J. (1978) *Motorways Versus Democracy*. Macmillan: London.

Utton, A. E., W. R. D. Sewell and T. O'Riordan (eds) (1976) *Natural Resources for a Democratic Society*. Westview: Boulder, Co.

Van Horn, C. E., Baumer, D. C. and W. T. Gormley (1989) *Politics and Public Policy*. CQ Press: Washington, DC.

Van Horn, C. E. and Y. Chilik (1988) How clean is clean? A case study of the nation's no. 1 Superfund toxic dump. *EIA Review*, **8** (2), 133–48.

Vari, A. (1989) Approaches towards conflict resolution in decision processes, in C. Vlek and G. Cvetkovich (eds) *Social Decision Methodology for Technological Projects*. Kluwer: Dordrecht, 79–84.

Vizayakumar, K. and P. K. J. Mohapatra (1991) Framework for environmental impact analysis – with special reference to India. *Environmental Management*, **15** (3), 357–68.

Wallerstein, I. (1979) *The Capitalist World Economy*. Cambridge University Press: Cambridge.

Walters, C. J. (1986) *Adaptive Management of Renewable Resources*. Macmillan: New York.

Wandesforde-Smith, G. (1989) Environmental impact assessment, entrepreneurship, and policy change, in R. V. Bartlett (ed) *Policy Through Impact Assessment*. Greenwood: Westport, Conn., 155–66.

Wandesforde-Smith, G. and J. Kerbavaz (1988) The co-evolution of politics and policy: elections, entrepreneurship and EIA in the United States, in P.Wathern (ed) *Environmental Impact Assessment: Theory and Practice*. Unwin Hyman: London, 161–91.

Wathern, P. (1988a) An introductory guide to EIA, in P.Wathern (ed) *Environmental Impact Assessment: Theory and Practice*. Unwin Hyman: London, 3–30.

Wathern, P. (ed) (1988b) *Environmental Impact Assessment: Theory and Practice*. Unwin Hyman: London.

Wathern, P. (1988c) Containing reform: the UK stance on the European Community EEC Directive. *Impact Assessment Bulletin*, **6** (3/4), 95–104.

Wathern, P. (1989) Implementing supranational policy: environmental impact assessment in the United Kingdom, in R. V. Bartlett, (ed) *Policy Through Impact Assessment*. Greenwood Press: Westport, Conn., 27–36.

Weary, G. C. (1984) Energy facility siting and routing, in Environment Canada *Facility Siting and Routing '84: Energy and Environment*, 2 vols. Supply and Services Canada: Ottawa, 1–12.

Weir, C. H. and J. P. Reay (1984) The corridor concept: theory and application, in Environment Canada *Facility Siting and Routing '84: Energy and Environment*, 2 vols. Supply and Services Canada: Ottawa, 121–52.

Wenner, L. M. (1989) The courts and environmental policy, in J. P. Lester (ed) *Environmental Politics and Policy*. Duke University Press: Durham, NC, 238–60.

Wenner, L. M. (1990) Environmental policy in the courts, in N. J. Vig and M. E. Kraft (eds) *Environmental Policy in the 1990s*. CQ Press: Washington, DC, 189–210.

Westman, W. E. (1985) *Ecology, Impact Assessment and Environmental Planning*. John Wiley: New York.

White, G. F. (1945) *The Human Adjustment to Floods*. Department of Geography Research Paper No. 29, University of Chicago: Chicago.

White, G. F. (1957) A perspective of river basin development. *Law and Contemporary Problems*, **22** (2), 157–84.

Whitney, J. B. R. (1985) Integrated economic–environmental models in environmental impact assessment, in J. B. R. Whitney and V. W. Maclaren (eds) *New Directions in Environmental Impact Assessment in Canada*. Methuen: Toronto, 53–86.

Whitney, J. B. R. and V. W. Maclaren (1985) A framework for the assessment of EIA methodologies, in J. B. R. Whitney and V. W. Maclaren (eds) *Environmental Impact Assessment: The Canadian Experience*. Institute for Environmental Studies, University of Toronto: Toronto, 1–31.

Whyte, A. V. and I. Burton (eds) (1980) *Environmental Risk Assessment*. John Wiley: Chichester.

Wildavsky, A. (1973) If planning is everything, maybe it's nothing. *Policy Sciences*, **4**, 127–53.

Wilson, D. C. and W. S. Forester (1987) Summary and analysis of hazardous waste management in ISWA countries, in W. S. Forester, and J. H. Skinner (eds) *International Perspectives on Hazardous Waste Management*. Academic Press: London, 11–96.

Wilson, V. S. (1981) *Canadian Public Policy and Administration*. McGraw-Hill Ryerson: Toronto.

Wood, C. J. B. (1976) Conflict in resource management and the use of threat: the Goldstream controversy. *Natural Resources Journal*, **16** (1), 137–58.

Wood, C. M. (1988) EIA in plan making, in P. Wathern (ed.) *Environmental Impact Assessment: Theory and Practice*. Unwin Hyman: London, 98–114.

Wood, C. M. (1989) *Planning Pollution Prevention*. Heinemann Newnes: Oxford.

Wood, C. M. and N. Lee (1988) The European directive on environmental impact assessment: implementation at last? *The Environmentalist*, **8** (3), 177–86.

Bibliography

Wood, C. M. and G. McDonic (1989) Environmental assessment: challenge and opportunity. *The Planner*, **75** (11), 12–18.

World Commission on Environment and Development (1987) *Our Common Future*. Oxford University Press: Oxford.

Wyatt, R. (1989) *Intelligent Planning*. Unwin Hyman: London.

Zallen, M. et al (1987) Follow-up review of impact assessments within the Coquihalla Valley, British Columbia, in B. Sadler (ed) *Audit and Evaluation in Environmental Assessment and Management*, vol.1. Supply and Services Canada: Ottawa, 122–44.

Index

Index

NET